Building Information Modeling and Geographic Information Systems for Next-Generation Spatial Intelligence

Mohamed Sobih Aly El Mekawy
https://orcid.org/0009-0006-3358-1368
Stockholm University, Sweden

IGI Global
Scientific Publishing
Publishing Tomorrow's Research Today

Vice President of Editorial	Melissa Wagner
Director of Acquisitions	Mikaela Felty
Director of Book Development	Jocelynn Hessler
Production Manager	Mike Brehm
Cover Design	Jose Rosado

Published in the United States of America by
IGI Global Scientific Publishing
701 East Chocolate Avenue
Hershey, PA, 17033, USA
Tel: 717-533-8845 | Fax: 717-533-7115
Website: https://www.igi-global.com E-mail: cust@igi-global.com

Library of Congress Cataloging-in-Publication Data

LCCN: 2025033232 (CIP Data Pending)
ISBN13: 9798337340586
Isbn13Softcover: 9798337340593
EISBN13: 9798337340609

British Cataloguing in Publication Data
A Cataloguing in Publication record for this book is available from the British Library.

Table of Contents

Detailed Table of Contents

Three-dimensional (3D) modeling and visualization have emerged as cornerstone technologies in a wide range of industries, transforming how we design, simulate, communicate, and interact with the world. From the early days of computer graphics to the cutting-edge realism of modern-day digital environments, the journey of 3D modeling and visualization is a fascinating tale of technological progress, creative ingenuity, and cross-disciplinary innovation. At its core, 3D modeling refers to the process of creating a mathematical representation of any three-dimensional object or surface using specialized software. Visualization, on the other hand, is the process of rendering these models into lifelike images or animations, often enhanced with textures, lighting, and shading to simulate reality or convey abstract concepts. The roots of 3D modeling trace back to the development of computer graphics in the mid-20th century. I

The integration of Building Information Modeling (BIM) and Geographic Information Systems (GIS) is transforming the landscape of urban planning by enabling intelligent, data-driven, and sustainable city development. This chapter explores the synergy of BIM-GIS convergence through the lens of Digital Twin applications, real-time simulation, and smart infrastructure planning. It discusses technological frameworks, standards, and visualization tools that support interoperability and spatial intelligence. Real-world case studies from Singapore and Helsinki demonstrate the practical benefits in underground utility mapping, participatory planning, and green zoning. The chapter also highlights emerging technologies such as IoT, AI, and edge computing that enhance spatial intelligence. Legal, governance, and adoption challenges are critically examined, along with future research directions.

data governance, cross-agency interoperability, and their functionality as publicly accessible information. Best practice case studies by major cities like Singapore, Helsinki, and Rotterdam show the real results that can be achieved in streamlining permitting and public engagements as well as facilitating climate resilient plans. In the future, the chapter outlines the future trends in predictive zoning, IoT-informed adaptive control, blockchain-mediated workflow, and immersive citizen engagement tools. As a tool to bridge the gap between technology and policy, BIM-GIS integration is not just an upgrade to the operational level, but a facilitator of the next generation of spatial intelligence in regulatory and zoning governance-flexible to allow cities to move beyond reactive enforcement to proactive, participatory, and informed decision-making.

Sustainable urban design and management are essential to address rapid urbanization, climate change, and resource depletion. This chapter examines the integration of Building Information Modeling (BIM) and Geographic Information Systems (GIS) as a comprehensive approach to creating smarter, greener cities. BIM offers building-level precision, while GIS provides spatial analytics and large-scale environmental context. Their integration enables informed decision-making, efficient resource use, and resilient infrastructure development. The chapter outlines theoretical frameworks, advances in 3D modeling and data interoperability, and real-world applications in green infrastructure, environmental impact assessment, and sustainable land use planning. Emphasizing BIM–GIS's role in advancing sustainable design principles, it serves as a guide for researchers, practitioners, and policymakers to adopt integrated spatial intelligence tools for sustainable urban transformation.

Urban infrastructure is the backbone of modern cities, encompassing transportation systems, utilities, buildings, public spaces, and communication networks. As cities grow and become more complex, the need for intelligent, data-driven infrastructure planning and management becomes crucial. Two powerful technologies—Building Information Modeling (BIM) and Geographic Information Systems (GIS)—are revolutionizing the way urban infrastructure is designed, constructed, operated, and maintained. By integrating these systems, urban planners, engineers, architects, and

Preface

The rapid digital transformation of cities, infrastructure systems, and environmental management has placed spatial intelligence at the forefront of contemporary research and practice. Building Information Modeling (BIM) and Geographic Information Systems (GIS), once distinct domains serving different scales and objectives, are converging into a unified ecosystem capable of supporting next-generation decision-making. This edited volume, *Building Information Modeling and Geographic Information Systems for Next-Generation Spatial Intelligence*, brings together emerging perspectives, technological innovations, and practical applications that illustrate the profound impact of BIM–GIS integration on the built and natural environments.

Across disciplines, the fusion of geometric precision with geospatial context is reshaping how we plan, design, construct, assess, and manage cities. Digital Twins, City Information Modeling (CIM), advanced 3D visualization, and intelligent mobility analytics have become indispensable tools for addressing pressing global challenges from climate resilience and green infrastructure to equitable urban services and transparent regulatory systems. As urbanization intensifies and sustainability becomes non-negotiable, the need for interoperable, data-driven, and citizen-centric approaches grows ever more urgent.

This book brings together scholars and practitioners from India, Uzbekistan, Cameroon, Pakistan, Oman, Bangladesh, Sweden, and beyond. Their contributions reflect diversity not only in geography, but also in disciplinary perspective, spanning engineering, urban planning, computer science, environmental design, transportation systems, and governance. Our chapters collectively offer a roadmap for harnessing BIM–GIS synergy to build smarter, greener, and more resilient cities.

CHAPTER OVERVIEW

Chapter 1: Advances in 3D Modeling and Visualization Technologies

Selvakumar P

This opening chapter traces the evolution of 3D modeling and visualization from early computer graphics to today's highly realistic digital environments. It explains the fundamental principles of 3D modeling and rendering, detailing how lifelike visualizations support design, simulation, and communication across industries. By examining technological progress and multidisciplinary innovation, the chapter establishes the conceptual foundation for understanding BIM and GIS as part of the broader evolution of spatial technologies.

Chapter 2: Digital Twin Applications in Urban Planning: BIM and GIS Synergy for Real-Time City Simulation

Ravikumar R N & Aarthi S

Chapter 2 explores how the integration of BIM and GIS enables Digital Twin–based urban planning, offering dynamic insights into real-time city operations. The authors examine interoperability standards, visualization tools, and data frameworks that support advanced spatial intelligence. Illustrative case studies from Singapore and Helsinki demonstrate applications in underground utility mapping, participatory urban design, and sustainable zoning. The chapter also foregrounds emerging technologies such as IoT, AI, and edge computing, while critically reviewing governance and adoption challenges.

Chapter 3: Digital Twins for Green Infrastructure—Extending BIM and GIS Integration

Ulugbek Eshqarayev, Shokhjakhon Akhmedov, Baxtiyor Egamov & Munir Ahmad

Focusing on ecological resilience, this chapter examines how Digital Twins combining BIM and GIS strengthen the planning and management of green infrastructure such as wetlands, green roofs, and urban parks. The authors highlight the capacity of real-time monitoring and performance modeling to promote climate adaptation and sustainability. They discuss technical and institutional limitations that currently hinder widespread implementation, and offer strategic pathways for ethical, inclusive, and scalable green Digital Twin solutions.

Chapter 4: Leveraging City Information Modeling (CIM) and GIS for Sustainable Road Infrastructure: The SIC Housing Access Road in Mfou, Cameroon

Cyntia Kengne Ngonda Tassi, Bikie Gerald Anicet, Roger Blaise Limaleba & Yassine Mouniane

This chapter presents a comprehensive digital workflow for the development of a key access road supporting a major housing initiative in Cameroon. Using GIS, CIM, Autodesk InfraWorks, AutoCAD Civil 3D, and Primavera P6, the researchers create a multi-dimensional project model that integrates topography, utilities, geometry, schedules, and cost estimates. The case study illustrates improved accuracy, collaboration, and stakeholder engagement, demonstrating how BIM–GIS integration accelerates infrastructure development in emerging contexts.

Chapter 5: Leveraging GIS for Intelligent Traffic Flow and Public Transport Planning: GeoSmart Mobility

Sitesh Kumar Singh, Jaishri Gothania & Leevesh Kumar

Chapter 5 delves into the transformational role of GIS in modern mobility systems. It analyzes GIS applications in congestion assessment, multimodal route optimization, and spatial equity, drawing on global examples from Singapore, Amsterdam, India, and the United States. The authors address critical issues including data privacy, institutional constraints, interoperability, and the role of open data ecosystems. Their conclusion positions GIS as a cornerstone of resilient and inclusive future transportation networks.

Chapter 6: Regulatory and Zoning Applications

S. Ida Evangeline

This chapter investigates how BIM and GIS can modernize regulatory and zoning processes. Traditional static and manual systems are contrasted with dynamic, data-rich approaches that incorporate 3D visualization, automated compliance checks, and simulation of policy impacts. The chapter outlines the evolution of zoning frameworks, technical strategies for BIM–GIS integration, and the emerging use of Digital Twins in real-time governance. It highlights how these technologies can significantly enhance urban transparency, efficiency, and adaptability.

Chapter 7: Sustainable Urban Design and Management: The Role of BIM and GIS Integration

Jyoti Rani & Shashiranjan Yadav

The authors emphasize BIM–GIS integration as a foundational tool for sustainable urban transformation. Through theoretical frameworks and practical case applications, the chapter explores how combining detailed building information with large-scale environmental analytics supports green infrastructure, environmental impact assessment, and sustainable land-use planning. It offers guidance for researchers, practitioners, and policymakers on adopting integrated spatial intelligence in sustainable city design.

Chapter 8: Use of BIM and GIS in Maintenance and Operation of Urban Infrastructure

Selvakumar P

The final chapter addresses the ongoing operation and maintenance of urban infrastructure, highlighting how BIM and GIS together support lifecycle management across transport systems, utilities, buildings, and public spaces. The chapter explains how integrated data environments enable informed decision-making, optimized resource allocation, and enhanced sustainability. By examining 3D and data-rich representations of urban assets, the chapter underscores BIM–GIS integration as a transformative tool for long-term urban resilience.

In bringing together these diverse contributions, this volume underscores the transformative potential of integrating BIM and GIS to reshape the future of urban environments. The chapters collectively illustrate that next-generation spatial intelligence is not merely a technological shift but a paradigm change—one that demands interdisciplinary collaboration, ethical data practices, and a commitment to sustainable and inclusive development. It is our hope that this book serves as both a resource and an inspiration for researchers, practitioners, and policymakers seeking to harness digital innovation for smarter, greener, and more resilient cities.

Chapter 1
Advances in 3D Modeling and Visualization Technologies

P. Selvakumar
https://orcid.org/0000-0002-3650-4548
Department of Science and Humanities, Nehru Institute of Technology, Coimbatore, India

ABSTRACT

Three-dimensional (3D) modeling and visualization have emerged as cornerstone technologies in a wide range of industries, transforming how we design, simulate, communicate, and interact with the world. From the early days of computer graphics to the cutting-edge realism of modern-day digital environments, the journey of 3D modeling and visualization is a fascinating tale of technological progress, creative ingenuity, and cross-disciplinary innovation. At its core, 3D modeling refers to the process of creating a mathematical representation of any three-dimensional object or surface using specialized software. Visualization, on the other hand, is the process of rendering these models into lifelike images or animations, often enhanced with textures, lighting, and shading to simulate reality or convey abstract concepts. The roots of 3D modeling trace back to the development of computer graphics in the mid-20th century. I

INTRODUCTION TO 3D MODELING AND VISUALIZATION.

Three-dimensional (3D) modeling and visualization have emerged as cornerstone technologies in a wide range of industries, transforming how we design, simulate, communicate, and interact with the world. From the early days of computer graphics

DOI: 10.4018/979-8-3373-4058-6.ch001

to the cutting-edge realism of modern-day digital environments, the journey of 3D modeling and visualization is a fascinating tale of technological progress, creative ingenuity, and cross-disciplinary innovation. At its core, 3D modeling refers to the process of creating a mathematical representation of any three-dimensional object or surface using specialized software. Visualization, on the other hand, is the process of rendering these models into lifelike images or animations, often enhanced with textures, lighting, and shading to simulate reality or convey abstract concepts. The roots of 3D modeling trace back to the development of computer graphics in the mid-20th century. In the 1960s, early pioneers such as Ivan Sutherland—whose program Sketchpad laid the foundation for interactive computer graphics—began exploring how computers could be used to create visual representations of objects. These early systems were crude by today's standards, limited by the computational power and display technologies of their time, but they marked the beginning of a transformative journey. Over the next few decades, advances in hardware, software, and mathematical algorithms enabled increasingly sophisticated representations of objects in three dimensions. The development of geometric modeling techniques such as polygonal modeling, NURBS (Non-Uniform Rational B-Splines),

Applications of 3D Modeling and Visualization: Architecture, Engineering, and Beyond

Three-dimensional (3D) modeling and visualization have emerged as cornerstone technologies in a wide range of industries, transforming how we design, simulate, communicate, and interact with the world. (Aboulnaga M. et al. 2024) From the early days of computer graphics to the cutting-edge realism of modern-day digital environments, the journey of 3D modeling and visualization is a fascinating tale of technological progress, creative ingenuity, and cross-disciplinary innovation. At its core, 3D modeling refers to the process of creating a mathematical representation of any three-dimensional object or surface using specialized software. Visualization, on the other hand, is the process of rendering these models into lifelike images or animations, often enhanced with textures, lighting, and shading to simulate reality or convey abstract concepts. (Aboulnaga M. et al. 2024) The roots of 3D modeling trace back to the development of computer graphics in the mid-20th century. In the 1960s, early pioneers such as Ivan Sutherland—whose program Sketchpad laid the foundation for interactive computer graphics—began exploring how computers could be used to create visual representations of objects. These early systems were crude by today's standards, limited by the computational power and display technologies of their time, but they marked the beginning of a transformative journey. Over the next few decades, advances in hardware, software, and mathematical algorithms enabled increasingly sophisticated representations of objects in three dimensions.

The development of geometric modeling techniques such as polygonal modeling, NURBS (Non-Uniform Rational B-Splines), and subdivision surfaces provided artists and engineers with powerful tools for crafting detailed and realistic models.

Initially, 3D modeling and visualization found application mainly in scientific research and industrial design. In aerospace and automotive engineering, 3D models allowed for precise visualization of components, facilitating simulation, analysis, and prototyping. In medicine, the ability to reconstruct anatomical structures from imaging data opened new avenues for diagnostics, surgery planning, and medical education (Adeyemi A. B. et al. 2024) Landmark films like *Jurassic Park* and *Toy Story* showcased the potential of computer-generated imagery (CGI), bringing to life characters and environments that were previously unimaginable. As the technology matured, the tools for 3D modeling and visualization became more accessible and powerful. Software such as Autodesk 3ds Max, Blender, Maya, SolidWorks, and SketchUp democratized the field, enabling not only professionals but also hobbyists and independent creators to experiment with 3D content creation. Parallel to this, the rise of powerful graphics processing units (GPUs) and real-time rendering engines like Unreal Engine and Unity revolutionized visualization capabilities, allowing for interactive experiences and immersive simulations in real-time. The convergence of these developments catalyzed the expansion of 3D applications into virtual reality (VR), augmented reality (AR), architecture, education, marketing, and beyond. A critical milestone in the evolution of 3D modeling has been the shift from static representations to dynamic, interactive, and intelligent models. Parametric modeling and procedural generation introduced levels of automation and adaptability, enabling the creation of complex structures with rules and logic rather than manual detailing. (Adeyemi A. B. et al. 2024) In architectural design, this allowed for building information modeling (BIM), a holistic approach that integrates geometry, spatial relationships, and construction data into a unified digital model. In gaming and simulation, procedural tools allow the generation of vast environments with high variability and realism, significantly reducing manual workload while enhancing scalability.

The Impact of 3D Modeling and Visualization on Industries and Society

The applications of 3D modeling and visualization span a vast array of industries, transforming workflows, improving communication, and enhancing the precision and creativity of design processes. (Alharbi A. A. 2025) At its core, 3D modeling enables professionals to create digital representations of physical objects or environments, while visualization brings these models to life through rendering techniques that simulate light, materials, and spatial relationships. Together, these technologies form a powerful toolkit used in architecture, engineering, construction, manufacturing, medicine, entertainment, and many other sectors. Their widespread adoption is driven by the ability to test ideas virtually, reduce physical prototyping costs, identify design flaws early, and communicate complex concepts in a visually intuitive manner. In architecture, 3D modeling has revolutionized how buildings and urban environments are conceived, presented, and constructed. Traditional 2D blueprints are now complemented—and often replaced—by detailed 3D models that allow architects to explore form, scale, and aesthetics with unprecedented depth. (Alharbi A. A. 2025) These models enable precise visualization of spatial relationships, materials, lighting, and environmental conditions, enhancing both design quality and client communication. Architects use software such as Autodesk Revit, Rhino, and SketchUp to develop models that can be easily modified and analyzed. Moreover, 3D visualization plays a critical role in project approvals, stakeholder engagement, and marketing by offering photorealistic renderings, walkthrough animations, and virtual reality (VR) experiences that help stakeholders fully understand the intent and impact of a design.

In the realm of engineering, 3D modeling and visualization are essential for designing and analyzing complex systems. (Anastasovitis E. et al. 2024) Mechanical engineers use computer-aided design (CAD) tools like SolidWorks, CATIA, and Siemens NX to create precise digital representations of mechanical parts and assemblies. These models can be subjected to simulations such as finite element analysis (FEA) and computational fluid dynamics (CFD) to evaluate structural integrity, thermal performance, and aerodynamics, long before a physical prototype is built. In civil and structural engineering, 3D models help visualize and analyze infrastructure projects such as bridges, tunnels, and skyscrapers, supporting collaboration between multidisciplinary teams and enhancing construction planning. (Anastasovitis E. et al. 2024) The integration of 3D modeling with Building Information Modeling (BIM) platforms has further elevated engineering workflows by linking geometry with data on materials, costs, scheduling, and maintenance, resulting in smarter, more efficient infrastructure. The construction industry has also embraced 3D technologies to reduce costs, avoid delays, and improve accuracy. Digital construction models allow contractors to plan site logistics, identify potential clashes between structural elements and systems (clash detection), and visualize construction sequencing. Technologies like 4D BIM (which integrates time as a dimension) and 5D BIM (which includes cost data) enable comprehensive project simulations, giving project managers a clearer understanding of how decisions affect timelines and budgets. Additionally, 3D laser scanning and drone photogrammetry are used to capture as-built conditions, monitor progress, and verify construction accuracy against design models, streamlining quality control and documentation.

Photogrammetry and 3D Scanning: Capturing Real-World Objects and Environments

The emergence and continual advancement of 3D modeling and visualization technologies have left a profound and far-reaching impact on both industries and society at large. Originally developed as niche tools for technical design and academic research, 3D modeling and visualization have now become pervasive, driving transformation in how businesses operate, how professionals communicate, how products are developed, and how people engage with the world. (Bhatt S. et al. 2025) The influence of these technologies can be seen in economic efficiencies, innovation acceleration, improved decision-making, enriched user experiences, and the democratization of creative expression. From an industrial and economic standpoint, 3D modeling and visualization have significantly streamlined workflows across sectors such as architecture, engineering, manufacturing, healthcare, entertainment, education, and marketing. (Bhatt S. et al. 2025) By enabling professionals to design, simulate, and iterate in a virtual environment, these tools reduce the need for costly physical prototypes, minimize errors, and shorten development cycles. In architecture and construction, for instance, the integration of 3D modeling with Building Information Modeling (BIM) platforms allows for the seamless collaboration of multidisciplinary teams, early identification of conflicts in designs, and more efficient construction planning. This leads to reduced project costs, fewer delays, and a more sustainable use of resources.

In engineering and manufacturing, the digital twin concept—creating a virtual replica of a physical product or system—has become a game-changer. (Chen Z. et al. 2024) Engineers can now monitor real-time performance, conduct simulations under various conditions, and predict failures before they occur, improving both product quality and operational reliability. The aerospace and automotive industries, in particular, have embraced these technologies to enhance safety, performance, and efficiency. By enabling precise modeling of complex geometries and materials, coupled with simulation of stress, fluid flow, and thermal dynamics, manufacturers can produce better designs while optimizing production processes. The healthcare sector has also experienced transformative change through 3D modeling and visualization. (Chen Z. et al. 2024) Clinicians use these technologies to reconstruct accurate anatomical models from medical imaging data for diagnostics, surgical planning, and custom prosthetic design. Surgeons can rehearse complex procedures using virtual simulations, increasing precision and reducing patient risk. Patients benefit from better communication, as visual models help them understand their conditions and treatment options more clearly. 3D printing, another offshoot of modeling, allows for the creation of patient-specific implants and tissue scaffolds, signaling a future of personalized medicine and regenerative treatments.

Computer-Aided Design (CAD): Parametric and Generative Modeling

Photogrammetry and 3D scanning represent two of the most powerful and widely adopted techniques for digitizing real-world objects and environments into accurate and detailed three-dimensional models. These technologies form the backbone of modern workflows in industries such as architecture, archaeology, engineering, film, gaming, forensics, cultural preservation, and virtual reality. (Chib S. et al. 2025) By enabling the capture of complex physical structures with millimeter or even sub-millimeter accuracy, photogrammetry and 3D scanning provide a critical bridge between the physical and digital realms, enhancing visualization, analysis, replication, and archiving. Photogrammetry is the process of extracting 3D information from 2D photographs. It is based on the principle of triangulation, where multiple overlapping images taken from different angles are used to determine the spatial position of points in three-dimensional space. Specialized software analyzes these images to identify common features and reconstruct a point cloud or mesh representing the object's shape and texture. One of photogrammetry's greatest advantages is its accessibility: it requires only a camera—ranging from a DSLR to a smartphone—and software tools like Agisoft Metashape, RealityCapture, or open-source alternatives such as Meshroom. (Chib S. et al. 2025) This democratization of 3D capture has made photogrammetry especially popular among independent creators, hobbyists, and researchers who need a low-cost, scalable solution for digitizing environments or objects.

On the other hand, 3D scanning encompasses a family of technologies that use sensors to directly measure the geometry of physical surfaces. (Dahiya R. et al. 2025) This includes laser scanning (LiDAR), structured light scanning, and time-of-flight scanning. Laser scanners, for example, emit laser beams that bounce back from the target object, measuring the time it takes for the light to return and thereby calculating distances. Structured light scanners project a known pattern (like stripes or grids) onto the object and analyze the deformation of the pattern to determine depth. (Dahiya R. et al. 2025) These methods generate highly detailed point clouds that can be converted into meshes and refined into watertight 3D models. Unlike photogrammetry, which heavily depends on image quality and lighting conditions, 3D scanning often delivers superior accuracy and is less sensitive to surface texture or lighting variability. A key difference between the two lies in data acquisition and processing complexity. Photogrammetry generally requires more intensive computational processing due to the need to analyze and align hundreds or even thousands of high-resolution images. It also requires careful planning of camera positions and overlap to ensure complete coverage of the subject. In contrast, many 3D scanners capture dense point clouds almost instantaneously and are particularly effective for capturing minute surface details, making them ideal for industrial inspection, reverse engineering, and quality control. However, scanners tend to be more expensive and may require calibration, controlled environments, and trained operators to function optimally.

3D Modeling Software: Advances in Tools and Techniques

Computer-Aided Design (CAD) has been a foundational technology in the digital transformation of design, engineering, and manufacturing industries. (Darsin M. et al. 2025) CAD refers to the use of computer systems to aid in the creation, modification, analysis, and optimization of a design. It has revolutionized how designers and engineers approach the conceptualization and realization of products, structures, and systems. Two of the most transformative methodologies within CAD are parametric modeling and generative modeling, each offering distinct advantages and reshaping workflows in fields ranging from mechanical engineering and architecture to industrial design and aerospace. Parametric modeling is a rule-based design approach where relationships between features, dimensions, and geometries are maintained through defined parameters and constraints. This technique allows for high levels of control, precision, and adaptability. In parametric systems, a model is not merely a static geometry—it is a dynamic construct that responds to input changes. For example, if a designer changes the diameter of a hole or the angle of a bracket, the entire model can update accordingly, preserving intended relationships and functional dependencies. This ability to make real-time, intelligent updates enables designers to iterate faster and accommodate late-stage changes without rebuilding the entire model. The logic behind parametric modeling lies in constraint-based design, where geometrical relationships (such as tangency, concentricity, perpendicularity, or equality) define how parts interact. (Darsin M. et al. 2025) Parameters—numerical values such as lengths, radii, or angles—drive the form of the object, making it easy to test variations and maintain consistency across families of parts. Tools like Autodesk Inventor, SolidWorks, CATIA, PTC Creo, and Fusion 360 are prime examples of parametric CAD platforms used across industries. This modeling approach is particularly suited to the development of mechanical components and assemblies, where dimensional fidelity, repeatability, and precision are critical.

In the architecture, engineering, and construction (AEC) sector, parametric modeling supports the creation of intelligent building components and complex geometries that conform to performance constraints. (García-Granero Á. et al. 2025) Building Information Modeling (BIM) platforms like Autodesk Revit employ parametric concepts, allowing users to model architectural elements (walls, windows, doors) with embedded data and behavior. This results in “smart” models that are not only visually accurate but also contain metadata related to materials, costs, thermal properties, and life cycle. (García-Granero Á. et al. 2025) Parametric BIM fosters collaboration, reduces errors, and facilitates design validation, cost estimation, and sustainability analysis throughout a building's lifecycle. Generative modeling, often associated with computational design and artificial intelligence, represents a significant leap beyond traditional parametric methods. Rather than defining every feature manually, generative modeling involves setting performance goals, design constraints, and material parameters, and then allowing the software to autonomously generate design solutions. It mimics natural processes such as evolution and growth to produce optimized forms, often resulting in organic, non-intuitive geometries that would be difficult to conceive through manual methods. Generative design is typically implemented through algorithms, scripting languages (like Grasshopper for Rhino), and AI-based platforms such as Autodesk's Generative Design tool, Bentley's Generative Components, or Siemens NX's topology optimization modules. These tools can produce thousands of design iterations based on structural efficiency, cost constraints, manufacturing methods, and sustainability goals. For example, in aerospace engineering, generative algorithms have created lightweight components that reduce fuel consumption without compromising strength, while in architecture, generative systems have helped design innovative facades that optimize sunlight exposure, airflow, and structural integrity.

Virtual Reality (VR) and Augmented Reality (AR): Immersive Visualization

The evolution of 3D modeling software has profoundly transformed the fields of design, engineering, architecture, entertainment, and beyond. Once the domain of highly specialized professionals and resource-intensive systems, 3D modeling software is now more powerful, accessible, and versatile than ever. (Giraldo J. P. et al. 2025) As technology advances, these tools have become indispensable in digital content creation and prototyping, driving innovation in industries as diverse as manufacturing, medicine, gaming, and education. From polygonal modeling to parametric and generative design, modern 3D software platforms offer a vast array of techniques that empower users to visualize, simulate, and fabricate with unprecedented precision and creativity. At its core, 3D modeling software is used to create digital representations of objects in three dimensions. These models serve as the foundation for everything from architectural blueprints and product designs to animated characters and immersive virtual worlds. (Giraldo J. P. et al. 2025) The earliest 3D modeling programs in the 1960s and 1970s were primarily developed for aerospace and defense, offering basic wireframe rendering and requiring expensive hardware. However, with the rise of personal computing in the 1980s and 1990s, software like AutoCAD, 3D Studio, and Alias brought 3D capabilities to a broader professional audience. Over time, these tools incorporated more advanced features such as NURBS (Non-Uniform Rational B-Splines), parametric modeling, real-time rendering, and scripting, marking a major leap forward in modeling efficiency and complexity.

Today's 3D modeling software landscape is both broad and specialized. (Gong Z. et al. 2024) Tools are generally categorized based on modeling techniques, application domains, and user needs. Polygonal modeling—widely used in gaming, film, and animation—relies on the manipulation of vertices, edges, and faces to form meshes. Programs like Autodesk Maya, Blender, and 3ds Max are staples in the entertainment industry, offering powerful sculpting, texturing, rigging, and animation capabilities. NURBS modeling, used in automotive and industrial design, allows for the creation of smooth, mathematically precise surfaces. Software like Rhino and Alias excels in this area, providing tools for both freeform and parametric surface modeling. Parametric and solid modeling are dominant in engineering and manufacturing. These rely on a history-based, feature-driven approach that defines geometry through parameters, constraints, and relations. (Gong Z. et al. 2024) Tools such as SolidWorks, Siemens NX, PTC Creo, and Autodesk Inventor allow for the creation of assemblies, simulation of mechanical movement, and integration with Computer-Aided Manufacturing (CAM) workflows. They are essential in product development cycles, offering tools for stress testing, motion analysis, and prototyping. Solid modeling ensures volumetric accuracy, making it ideal for parts intended for machining, 3D printing, or injection molding. Procedural and generative modeling represent cutting-edge developments in 3D software. Programs like SideFX Houdini use node-based workflows that enable users to create complex structures and effects through algorithms and logic-based operations. In architecture and computational design, tools like Grasshopper for Rhino or Autodesk Generative Design allow users to define rules and constraints, with the software generating optimized design solutions based on performance criteria. These tools are particularly well-suited to scenarios that require iterative exploration, performance-driven optimization, or the design of non-standard, organic geometries.

Real-Time Rendering: Fast and Efficient Visualization

Virtual Reality (VR) and Augmented Reality (AR) have emerged as groundbreaking technologies that redefine how we interact with digital content, offering immersive visualization experiences that are transforming numerous industries. (Hevko I. V. et al. 2025) By bridging the gap between the digital and physical worlds, VR and AR allow users to engage with three-dimensional models, simulations, and environments in ways that are more intuitive, spatially aware, and emotionally impactful than traditional screen-based interactions. As hardware becomes more affordable and software more sophisticated, the applications of VR and AR in design, education, healthcare, manufacturing, real estate, and entertainment are expanding rapidly, reshaping not only workflows but also human perception and behavior. Virtual Reality refers to fully immersive digital environments that completely replace the user's physical surroundings. Users wear head-mounted displays (HMDs) like the Meta Quest, HTC Vive, or PlayStation VR to enter these virtual spaces, often interacting with them through motion controllers or hand tracking. VR provides a stereoscopic 3D view, spatial audio, and often full-body tracking, creating a sense of presence—where users feel physically situated within a digital environment. This level of immersion is ideal for applications where total focus and spatial exploration are crucial, such as architectural walkthroughs, virtual training simulations, and virtual prototyping. In contrast, Augmented Reality overlays digital content onto the real world, enhancing, rather than replacing, the user's perception of their environment. (Hevko I. V. et al. 2025) AR is typically experienced through smartphones, tablets, or specialized AR headsets like Microsoft HoloLens or Magic Leap. Through sensors, cameras, and advanced computer vision algorithms, AR devices can recognize surfaces, track movements, and anchor digital content in real-world coordinates. This enables applications such as interactive manuals, product visualizations, medical diagnostics, and on-site construction aids. Because AR does not remove users from their environment, it is particularly well-suited to situations where real-time, contextual information enhances physical tasks or decision-making.

Both VR and AR rely heavily on 3D modeling and visualization. Every object, environment, or avatar presented in a VR or AR space must be designed and optimized for performance and realism. (Isikay I. et al. 2024) High-quality 3D assets, coupled with real-time rendering engines like Unity and Unreal Engine, allow developers to create richly detailed and interactive worlds. In VR, entire architectural spaces can be modeled to true scale, allowing architects, clients, and stakeholders to explore a building long before construction begins. (Zhou Y. et al. 2024) In AR, product designers can project virtual prototypes onto physical surfaces to assess scale, aesthetics, and ergonomics. These immersive experiences improve comprehension, foster engagement, and support iterative design with fewer physical prototypes. In architecture and construction, immersive visualization with VR enables full-scale walkthroughs of unbuilt spaces, where stakeholders can assess spatial relationships, lighting, material finishes, and layout in a way that 2D drawings or static renders cannot convey. (Zhou Y. et al. 2024) (Maharjan S. et al. 2024) Clients can virtually "walk" through their future homes or offices, providing feedback before ground is broken. In AR, construction professionals can overlay BIM data onto job sites to visualize where structural elements, utilities, or finishes will go—reducing errors, facilitating on-site decision-making, and improving communication among teams.

The automotive and manufacturing industries benefit significantly from VR and AR technologies. (Isikay I. et al. 2024) In VR, designers and engineers can collaborate remotely to review vehicle designs at full scale, analyze ergonomics, and test assembly sequences. VR simulators are also used for training factory workers and machine operators in safe, controlled environments. AR, on the other hand, is increasingly being used on factory floors, where technicians use AR headsets or mobile devices to receive step-by-step assembly instructions, diagnostics, or part identification in real time. This reduces reliance on paper manuals, speeds up repairs, and minimizes human error.

Data Visualization: Representing Complex Data in 3D

Real-time rendering has emerged as a cornerstone of modern digital visualization, revolutionizing how individuals and organizations interact with 3D environments and digital content. (Zheng X. et al. 2025) Unlike traditional offline rendering, which can take minutes or even hours per frame to calculate photorealistic visuals, real-time rendering enables the instantaneous generation of images at speeds fast enough to maintain fluid user interaction—typically 30 to 60 frames per second or more. (Jo H. et al. 2024) This technology has become essential in domains such as video games, virtual reality (VR), architectural visualization, product design, simulation training, and virtual prototyping, where responsiveness, interactivity, and visual fidelity are crucial. At its core, real-time rendering involves the process of calculating and displaying 3D scenes on-the-fly as users interact with them. (Zheng X. et al. 2025) It relies heavily on the power of modern Graphics Processing Units (GPUs), which are optimized for parallel processing and can handle millions of calculations per second. Real-time rendering engines, such as Unreal Engine, Unity, CryEngine, Godot, and proprietary in-house platforms, provide the backbone for creating interactive digital experiences. These engines integrate rendering algorithms with physics, lighting, animation, and AI systems to produce lifelike, responsive environments.

One of the primary innovations enabling real-time rendering is the rasterization process, where 3D geometry is projected onto a 2D screen and then shaded based on lighting models and texture maps. Rasterization is fast and efficient, making it suitable for real-time applications. Coupled with Level of Detail (LOD) algorithms, occlusion culling, and view frustum culling, real-time engines are able to render complex scenes by prioritizing what the camera can see and dynamically adjusting the resolution and detail of objects depending on their importance or distance. (Kaur M. et al. 2025) Real-time rendering has evolved significantly with the integration of Physically Based Rendering (PBR). PBR simulates the physical properties of light and materials to produce more realistic and consistent results under different lighting conditions. (Zhang L. et al. 2024) Materials created using PBR workflows respond accurately to environment lighting, reflections, and shadows, giving digital scenes a natural appearance. PBR has become the standard in real-time engines, enabling developers and designers to achieve cinematic quality in interactive experiences. (Kaur M. et al. 2025) In recent years, ray tracing, traditionally reserved for offline rendering due to its computational demands, has made its way into real-time workflows thanks to hardware acceleration from GPUs such as NVIDIA's RTX series and AMD's Radeon RX series. Real-time ray tracing allows for accurate reflections, refractions, global illumination, and soft shadows by simulating the path of light rays as they bounce around a scene. (Zhang L. et al. 2024) Though still more performance-intensive than rasterization, hybrid approaches combine rasterization for general scene rendering with ray tracing for select effects, striking a balance between realism and speed.

Architecture and Urban Planning: 3D Modeling and Visualization

Data visualization is a fundamental discipline in the era of big data, transforming raw numbers into visual formats that are more intuitive, interactive, and insightful. (Kommineni M. et al. 2025) With the exponential growth of data generated from scientific research, business analytics, engineering simulations, and IoT devices, the ability to understand and communicate complex datasets is more critical than ever. Traditional 2D charts and graphs—while still effective for many purposes—can become limited when dealing with multi-dimensional, dynamic, or spatially distributed information. This has driven the rise of 3D data visualization, an advanced approach that uses three-dimensional space to represent data structures, relationships, and patterns more comprehensively. At its core, 3D data visualization leverages geometric representations (such as points, lines, surfaces, and volumes) in a spatial coordinate system (X, Y, and Z axes) to reveal relationships in data that are not easily discernible in 2D. It adds depth and perspective to visualizations, offering new dimensions of interactivity and analysis. (Kommineni M. et al. 2025) This approach is especially valuable for fields where data is inherently spatial—such as geoscience, medical imaging, engineering simulations, molecular biology, and urban planning—but is also increasingly applied in abstract data contexts, like finance, machine learning, and network analysis, to better understand multi-variable interactions.

Scientific visualization is one of the earliest and most prominent domains to adopt 3D data visualization. (Lefranc G. et al. 2025) It involves rendering scientific data—often derived from simulations or sensor networks—in a visual form that researchers can study and manipulate. For example, in fluid dynamics, Computational Fluid Dynamics (CFD) simulations use 3D vector fields to show flow patterns around objects like aircraft or turbines. Weather models use volumetric renderings of temperature, pressure, and humidity fields to simulate storm systems. In astrophysics, simulations of galactic formation or black hole dynamics are visualized in 3D to reveal structural behavior and emergent patterns that cannot be easily understood through raw equations alone. (Zaman N. et al. 2024) In medical and biomedical visualization, 3D techniques are indispensable. Technologies such as CT (Computed Tomography), MRI (Magnetic Resonance Imaging), and PET (Positron Emission Tomography) generate volumetric data that must be rendered in 3D to analyze human anatomy and diagnose medical conditions. (Lefranc G. et al. 2025) These visualizations allow physicians and surgeons to explore internal structures—organs, tissues, blood vessels—in a virtual space, aiding in diagnostics, surgical planning, and education. (Zaman N. et al. 2024) In molecular biology, 3D visualization tools like PyMOL and Chimera allow scientists to examine protein structures, DNA strands, and molecular interactions in detail, which is critical for drug discovery and genetic engineering.

Product Design and Manufacturing: 3D Modeling and Prototyping

The integration of 3D modeling and visualization into architecture and urban planning has fundamentally transformed how designers conceptualize, communicate, and execute their ideas. (Li X. et al. 2025) From initial sketches to final construction, 3D technologies provide a powerful medium through which spatial ideas are developed, refined, and shared with clients, communities, and stakeholders. Once limited to two-dimensional blueprints, sections, and elevations, the architectural process now thrives in an immersive digital environment where buildings and cities are conceived with unprecedented depth, realism, and interactivity. This shift not only enhances aesthetic and technical quality but also facilitates collaboration, sustainability, and public engagement in the planning process. In architecture, 3D modeling begins in the early design phase, where conceptual massing studies are used to explore form, orientation, and spatial relationships. (Li X. et al. 2025) Digital tools like SketchUp, Rhino, and Autodesk Revit allow architects to quickly generate volumetric representations of structures, enabling real-time feedback and design iteration. These models can be manipulated, sliced, and viewed from multiple angles, offering a spatial clarity that 2D drawings cannot provide. (Yiğit A. Y. et al. 2025) More advanced parametric tools, such as Grasshopper for Rhino and Dynamo for Revit, allow for algorithmic control over geometry, enabling the creation of complex forms based on rules, constraints, and environmental factors.

Figure 1. Advanced Techniques in 3d Modeling

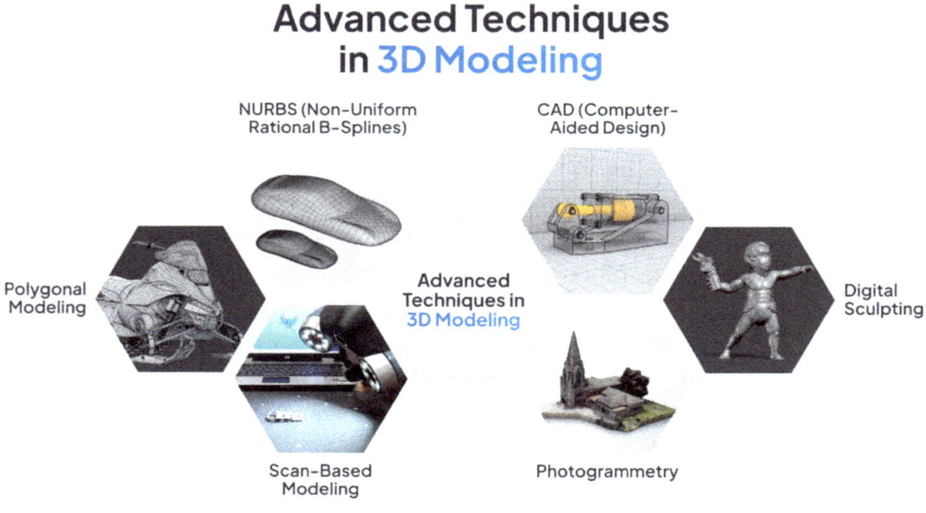

Building Information Modeling (BIM) has become a cornerstone in architectural practice, offering a data-rich 3D environment that integrates geometry with construction metadata. BIM platforms, particularly Autodesk Revit, Graphisoft ArchiCAD, and Bentley Systems, provide detailed information about materials, structural systems, energy performance, and lifecycle costs. (Limbore N. V. et al. 2025) Through BIM, architects, engineers, contractors, and clients can collaborate on a centralized digital model that ensures consistency, reduces errors, and supports informed decision-making. BIM models are not only visual representations—they are digital twins of buildings, supporting construction sequencing, clash detection, cost estimation, and facility management long after the design phase. (Limbore N. V. et al. 2025) 3D visualization technologies also play a vital role in architectural communication and client engagement. Rendering engines like Lumion, Twinmotion, V-Ray, and Enscape bring architectural models to life with photorealistic textures, lighting, vegetation, people, and atmospheric effects. These visualizations allow clients to virtually experience a building before it is constructed, enabling a deeper emotional and cognitive connection to the design. Virtual Reality (VR) further enhances this experience by allowing users to immerse themselves in a building at full-scale walk-through rooms, and understand spatial qualities in ways that traditional drawings cannot convey. (Yiğit A. Y. et al. 2025) This helps mitigate misunderstandings, accelerates approvals, and empowers clients to make more confident decisions.

Medical Visualization: 3D Modeling and Simulation: Overview of medical visualization.

3D modeling and prototyping have revolutionized the field of product design and manufacturing, enabling faster, more efficient, and more innovative development cycles than ever before. (Liu Z. et al. 2025) These technologies have transformed how products are conceptualized, tested, refined, and produced across industries ranging from consumer electronics and automotive to healthcare, aerospace, and industrial equipment. By leveraging digital 3D tools, designers and engineers can create, visualize, iterate, and validate their ideas with high precision and at significantly lower costs and timeframes than traditional approaches. At the heart of product design lies the use of 3D Computer-Aided Design (CAD) software. These tools allow engineers and designers to digitally construct precise geometric models of products or components, complete with all necessary details related to dimensions, materials, tolerances, and assembly. Programs like SolidWorks, Autodesk Fusion 360, PTC Creo, Siemens NX, and CATIA are among the most widely used platforms. They support both solid and surface modeling techniques and are capable of building complex parametric models that adapt automatically to design changes, greatly enhancing flexibility during development. The modeling phase enables conceptualization and form-finding. Industrial designers often begin with rough 3D sketches and digital sculpting tools (like Rhinoceros, Alias, or Blender) to explore aesthetics and ergonomics. (Liu Z. et al. 2025) These early-stage models help visualize how a product will look and feel, serving as the foundation for both creative exploration and stakeholder communication. From there, models evolve into detailed engineering representations with functional elements—such as joints, fasteners, and internal mechanisms—fully defined. The ability to simulate the fit, function, and motion of components within assemblies ensures that design flaws are identified and corrected early in the process.

One of the most transformative aspects of modern product development is virtual prototyping, which allows teams to simulate how a product will behave under real-world conditions—without having to build a physical prototype. (Maharjan S. et al. 2024) Tools such as ANSYS, COMSOL Multiphysics, and Autodesk Inventor Nastran offer advanced capabilities for Finite Element Analysis (FEA), Computational Fluid Dynamics (CFD), thermal simulations, vibration analysis, and other performance evaluations. These simulations allow for rigorous stress testing, material optimization, and energy efficiency assessment, enabling engineers to refine designs before any physical manufacturing occurs. In tandem with virtual prototyping, 3D rendering and visualization software such as KeyShot, VRED, and Luxion help designers produce photorealistic images and animations of their models. These visuals are not only useful for internal review and marketing but also for communicating ideas to non-technical stakeholders, clients, or investors. (Maharjan S. et al. 2024) The realism and clarity offered by advanced rendering tools facilitate better design decisions, enhance team collaboration, and often accelerate the approval process. (Xu H. et al. 2024) While virtual tools are powerful, the transition from digital to physical remains crucial—this is where rapid prototyping enters the workflow. Rapid prototyping involves quickly fabricating a scale model or functional version of a product using technologies like 3D printing, CNC machining, laser cutting, and vacuum casting. 3D printing, in particular, has become a cornerstone of modern prototyping due to its flexibility, speed, and cost-effectiveness. (Xu H. et al. 2024) It allows for the creation of complex geometries, internal structures, and moving parts that would be expensive or impossible to produce through traditional methods.

Artificial Intelligence (AI) in 3D Modeling and Visualization

Medical visualization, driven by advancements in 3D modeling and simulation, has become a transformative force in modern healthcare. (Maskeliūnas R. et al. 2025) By converting complex medical data into interactive three-dimensional representations, healthcare professionals can now better understand human anatomy, diagnose diseases more accurately, plan surgical procedures, and communicate with patients and colleagues in unprecedented ways. (Wu T. et al. 2024) The integration of 3D visualization into clinical practice, education, and biomedical research has opened new frontiers in precision medicine, minimally invasive surgery, and personalized treatment, fundamentally improving the quality, safety, and efficiency of care delivery. (Wu T. et al. 2024) At the core of medical visualization is the transformation of raw medical imaging data—such as CT (Computed Tomography), MRI (Magnetic Resonance Imaging), PET (Positron Emission Tomography), and ultrasound—into detailed 3D anatomical models. These imaging modalities produce volumetric data, which can be segmented and reconstructed into three-dimensional structures using specialized software such as 3D Slicer, OsiriX, Mimics, Amira, and Zebra Medical Vision. (Maskeliūnas R. et al. 2025) Through segmentation, different tissues (e.g., bones, organs, tumors, blood vessels) are identified and digitally isolated, allowing clinicians to create accurate, patient-specific models that reveal internal anatomy with remarkable clarity.

These 3D reconstructions serve a wide variety of clinical applications. In preoperative planning, for instance, surgeons use 3D models to assess a patient's unique anatomy before performing complex procedures. (Qiu S. et al. 2025) For neurosurgery, orthopedics, and cardiovascular surgery, understanding the spatial relationship between critical structures—such as arteries, nerves, or tumor margins—can mean the difference between success and serious complications. 3D visualization allows surgeons to virtually explore and rehearse operations, plan incisions, identify risks, and determine the optimal surgical path, all before entering the operating room. In minimally invasive and robotic surgery, 3D simulation plays an even more vital role. (Qiu S. et al. 2025) Surgeons can operate with enhanced precision using real-time 3D navigation systems that integrate preoperative models with intraoperative imaging. These systems guide instruments with sub-millimeter accuracy, helping to avoid vital structures and reduce trauma. Platforms like Medtronic's StealthStation, Intuitive Surgical's da Vinci, and Brainlab's Curve Navigation rely on 3D visualization to enhance control and situational awareness, leading to shorter recovery times and better patient outcomes.

Cloud-Based 3D Modeling and Visualization: Collaborative Tools

Artificial Intelligence (AI) is profoundly reshaping the landscape of 3D modeling and visualization, revolutionizing how digital content is created, refined, and interpreted. Traditionally, 3D modeling has relied heavily on manual effort—designers meticulously crafting geometric shapes, animators rigging characters frame-by-frame, and engineers running complex simulations. (Sachdeva R. et al. 2024) Today, AI is augmenting and automating many of these processes, dramatically accelerating workflows, enhancing realism, enabling personalization, and opening new frontiers in interactivity and computational creativity. From generative design and content synthesis to semantic segmentation and intelligent rendering, AI is becoming an indispensable tool in industries ranging from gaming and film to architecture, medicine, manufacturing, and scientific visualization. (Wang P. et al. 2025) One of the most transformative areas is AI-assisted content creation, where machine learning models help generate 3D models, textures, and scenes with minimal user input. Tools powered by Generative Adversarial Networks (GANs) and Variational Autoencoders (VAEs) can synthesize realistic 3D objects or terrains from sketches, textual descriptions, or even photographs. (Wang P. et al. 2025) For example, NVIDIA's GauGAN and GET3D models can turn 2D images into detailed 3D environments or create high-fidelity 3D meshes with learned topology and textures. Similarly, OpenAI's Point-E and Google's DreamFusion generate 3D objects directly from text prompts using diffusion models and neural rendering techniques. (Sachdeva R. et al. 2024) These technologies empower creators with new levels of abstraction, enabling them to iterate rapidly and focus on high-level design rather than low-level geometry.

In the realm of generative design, AI algorithms explore vast design spaces to discover optimal shapes, (Saifi S. et al. 2025) structures, and configurations based on defined constraints and objectives. In engineering and product design, software such as Autodesk Fusion 360 with Generative Design, Siemens NX, and nTopology use AI to propose lightweight, high-performance geometries that human designers might not conceive. These AI-generated forms are often organically shaped and structurally efficient, ideally suited for additive manufacturing. (Saifi S. et al. 2025) By integrating performance simulation and cost analysis into the design loop, AI ensures not only aesthetic innovation but also material efficiency, functional integrity, and manufacturability. AI also plays a pivotal role in 3D reconstruction and scene understanding. Using data from multiple images or depth scans, AI models can infer complete 3D geometry and structure. Computer vision techniques such as multi-view stereo, structure from motion, and monocular depth estimation are enhanced with deep learning to produce more accurate and detailed reconstructions. (Vaidyanathan B. et al. 2025) For instance, AI-powered photogrammetry tools can automatically generate 3D models of real-world objects or environments from photos or video, streamlining workflows in cultural heritage preservation, gaming, and virtual production. (Vaidyanathan B. et al. 2025) In autonomous vehicles and robotics, semantic 3D scene understanding—powered by convolutional neural networks (CNNs) and transformers—enables machines to recognize and navigate through their environments in real time.

Future Trends: Emerging Technologies and Applications

Cloud-based 3D modeling and visualization platforms have emerged as a transformative force in digital design, engineering, and creative industries, redefining how teams collaborate, iterate, and deploy 3D content. (Samal A. et al. 2025) Traditionally, 3D modeling workflows were constrained to high-performance desktop environments, tethered to local software licenses, rigid file structures, and hardware limitations. (Ullah M. et al. 2024) As cloud computing has evolved, so too has the ability to shift complex modeling, rendering, and simulation processes onto distributed infrastructures, enabling real-time collaboration, scalability, and seamless integration across geographies and disciplines. The rise of cloud-native tools and platforms is fundamentally reshaping not only the technical landscape of 3D design but also the cultural and organizational dynamics of how work is done. (Ullah M. et al. 2024) Cloud-based 3D tools allow users to access powerful modeling and visualization capabilities through a web browser or lightweight client, eliminating the need for heavy installations or expensive workstations. These platforms leverage the computing power of centralized data centers to perform demanding tasks such as rendering, simulation, and file synchronization. (Samal A. et al. 2025) Leading solutions include Autodesk Fusion 360, Onshape, Trimble SketchUp for Web, Siemens NX X, Clara.io, Shapr3D Cloud, and Unity Reflect. These tools are designed to run efficiently on cloud infrastructure, supporting a wide range of use cases including CAD design, product prototyping, architectural visualization, animation, and interactive simulations.

One of the key advantages of cloud-based modeling is real-time collaboration. Unlike traditional workflows, (Shaikh M. et al. 2024) where models are shared via email or FTP and require manual merging of changes, cloud platforms enable multiple users to work on the same model simultaneously. This live collaboration is facilitated through version control systems, granular permissions, and activity tracking—similar to how Google Docs revolutionized document editing. (Uddin M. K. S. et al. 2024) For example, in Onshape, users can co-edit assemblies in real time, annotate parts, and instantly see updates without the need to resend files. This collaborative model reduces version conflicts, accelerates iteration cycles, and empowers cross-functional teams to work more cohesively. (Shaikh M. et al. 2024) In industries like architecture, engineering, and construction (AEC), where collaboration across disciplines is essential, cloud-based Building Information Modeling (BIM) tools such as Autodesk BIM 360, Trimble Connect, and Revizto offer centralized environments for model sharing, coordination, and clash detection. These platforms support complex federated models, where architectural, structural, and MEP (mechanical, electrical, plumbing) elements from different teams can be merged, reviewed, and resolved collaboratively. (Uddin M. K. S. et al. 2024) Cloud integration ensures that all project stakeholders—from designers and contractors to clients and regulators—are working with the most current data, reducing errors, miscommunication, and costly rework.

The Future of 3D Modeling and Visualization: Opportunities and Challenges

The future of 3D modeling and visualization is poised to be profoundly shaped by a convergence of cutting-edge technologies, interdisciplinary innovations, and evolving user demands that together promise to redefine how digital content is created, interacted with, and applied across diverse sectors. (Sharma A. et al. 2024) As we move further into the 21st century, rapid advancements in Artificial Intelligence (AI), Extended Reality (XR), quantum computing, edge computing, cloud infrastructure, and materials science are driving a new era of immersive, intelligent, and interactive 3D experiences. These emerging trends are not only enhancing existing workflows but are also enabling entirely novel applications that could transform industries ranging from entertainment and healthcare to manufacturing, education, urban planning, and beyond. One of the most transformative drivers of future progress is the integration of Artificial Intelligence and Machine Learning into virtually every stage of the 3D pipeline. (Sharma A. et al. 2024) Beyond current applications in automated segmentation, generative design, and texture synthesis, future AI models will evolve into more autonomous creative partners. (Sreekanthaswamy N. et al. 2025) Advances in generative adversarial networks (GANs), diffusion models, and reinforcement learning will empower AI to conceptualize, generate, and optimize complex 3D assets from minimal input—whether textual, conceptual, or procedural—with unprecedented fidelity and creativity. This "AI as co-creator" paradigm will streamline workflows, democratize design for non-experts, and accelerate iteration cycles, while fostering new styles and forms that challenge traditional artistic boundaries.

Simultaneously, Extended Reality (XR)—encompassing Virtual Reality (VR), Augmented Reality (AR), and Mixed Reality (MR)—is expected to become the primary medium through which users engage with 3D content. (Soundariya A. et al. 2025) Unlike conventional screen-based visualization, XR enables fully immersive and interactive experiences that blend physical and digital realities. (Sreekanthaswamy N. et al. 2025) Future XR platforms will leverage advances in lightweight, high-resolution headsets, haptic feedback devices, eye tracking, and spatial audio to deliver hyper-realistic environments for design review, training simulations, collaborative prototyping, and entertainment. (Tamilarasi C. et al. 2025) The ability to manipulate digital twins, architectural models, or anatomical visualizations in three-dimensional space, while sharing these environments with remote collaborators in real time, will revolutionize remote work and learning paradigms. A related development lies in the rise of digital twins—dynamic, real-time virtual replicas of physical systems, processes, or environments. (Soundariya A. et al. 2025) Powered by IoT (Internet of Things) sensors, AI analytics, and cloud computing, digital twins will enable continuous monitoring, simulation, and predictive maintenance of complex assets such as smart cities, industrial machinery, healthcare systems, and supply chains. In urban planning, for instance, digital twins will allow planners to simulate traffic flow, energy consumption, and environmental impact with high spatial and temporal resolution, facilitating more sustainable and resilient development. (Tamilarasi C. et al. 2025) The fidelity and responsiveness of these digital twins will depend heavily on advancements in real-time 3D visualization and simulation technologies, alongside improved data integration and interoperability standards.

CONCLUSION

The future of 3D modeling and visualization stands at the intersection of immense opportunities and complex challenges, shaped by rapid technological advancements and evolving user expectations. As digital creation and representation technologies continue to mature, the potential to transform industries such as architecture, manufacturing, entertainment, healthcare, education, and urban planning grows exponentially. These advancements promise to not only enhance the quality, speed, and accessibility of 3D content creation but also to redefine the way people interact with digital environments. However, alongside these exciting prospects, the future landscape presents significant hurdles related to technology integration, ethical considerations, infrastructure demands, and workforce adaptation that must be addressed to fully realize the benefits. One of the most prominent opportunities lies in the integration of Artificial Intelligence (AI) and machine learning into 3D workflows, which is poised to revolutionize content creation, optimization, and visualization. AI-powered tools will increasingly automate repetitive and complex tasks such as mesh generation, texture synthesis, animation rigging, and predictive simulation, enabling designers and engineers to focus more on creative and strategic aspects. Generative design algorithms will explore vast solution spaces to optimize form, function, and material usage, especially in engineering and product design, leading to innovative and sustainable solutions. In addition, AI's ability to analyze large datasets will enhance 3D reconstructions from images or scans, making it easier to create accurate digital twins of physical objects and environments for simulation, monitoring, and planning purposes.The rise of cloud computing and cloud-based collaborative platforms will continue to democratize access to powerful 3D modeling and visualization tools. Cloud infrastructure allows for real-time multi-user collaboration, seamless version control, and remote access, breaking down geographical and technical barriers. This is particularly transformative in scenarios where distributed teams, clients, and stakeholders must interact with shared digital assets, such as in architectural design or complex manufacturing projects. Cloud platforms also facilitate scalable rendering and simulation capabilities, allowing users to handle computationally intensive tasks without investing in costly hardware. This shift lowers the entry barrier for startups, educational institutions, and individual creators, fostering innovation and inclusivity. Extended Reality (XR) technologies—including Virtual Reality (VR), Augmented Reality (AR), and Mixed Reality (MR)—are rapidly becoming pivotal in how 3D content is consumed and manipulated. The future of 3D visualization is increasingly immersive, enabling users to experience spatial models in lifelike environments, interact with digital content intuitively, and collaborate in virtual shared spaces regardless of physical location. This holds significant promise across fields such as medical training, where realistic anatomical models can improve surgical preparation; urban planning, where city models can be explored at scale; and entertainment, where interactive narratives and gaming experiences reach new levels of engagement. The convergence of AI and XR will further personalize experiences, adapting virtual environments in real time to user preferences and behavior.

REFERENCES

Aboulnaga, M., Abouaiana, A., Puma, P., Elsharkawy, M., Farid, M., Gamal, S., & Lucchi, E. (2024). Climate change and cultural heritage: A global mapping of the UNESCO thematic indicators in conjunction with advanced technologies for cultural sustainability. *Sustainability (Basel)*, *16*(11), 4650. DOI: 10.3390/su16114650

Adeyemi, A. B., Ohakawa, T. C., Okwandu, A. C., Iwuanyanwu, O., & Ifechukwu, G. O. (2024). Advanced Building Information Modeling (BIM) for affordable housing projects: Enhancing design efficiency and cost management. *Journal of Building Information Modeling*, *12*(1), 45–60.

Alharbi, A. A. (2025). Cognitive learning approach to enhance university students' visualization of molecular geometry in chemical compounds: A case study in Saudi Arabia. *Journal of Radiation Research and Applied Sciences*, *18*(1), 101283. DOI: 10.1016/j.jrras.2024.101283

Anastasovitis, E., Georgiou, G., Matinopoulou, E., Nikolopoulos, S., Kompatsiaris, I., & Roumeliotis, M. (2024). Enhanced inclusion through advanced immersion in cultural heritage: A holistic framework in virtual museology. *Electronics (Basel)*, *13*(7), 1396. DOI: 10.3390/electronics13071396

Bhatt, S., Simre, S. S., Patadiya, H. H., Yadav, D., Deepti, B., Sathvi, S. S., & Manamasa, Y. (2025). 3D Printing in Dentistry: A Review. *Journal of Advanced Medical and Dental Sciences Research*, *13*(1), 11–15.

Chen, Z., Chen, L., Zhou, X., Huang, L., Sandanayake, M., & Yap, P. S. (2024). Recent technological advancements in BIM and LCA integration for sustainable construction: A review. *Sustainability (Basel)*, *16*(3), 1340. DOI: 10.3390/su16031340

Chib, S., Mehta, A. K., Selvakumar, P., Saravanan, A., Mishra, B. R., & Manjunath, T. C. (2025). Mental health challenges and solutions in high-pressure work environments. In Harriott, S. (Ed.), *Innovative approaches to managing conflict and change in diverse work environments* (pp. 247–272). IGI Global Scientific Publishing. DOI: 10.4018/979-8-3693-9556-1.ch011

Dahiya, R., & Selvakumar, P. Vidyadhara, & Manjunath, T. C. (2025). Design and aerodynamics: Principles and advances. In M. Mellal (Ed.), Innovations and developments in unmanned aerial vehicles (pp. 145-170). IGI Global Scientific Publishing.

Darsin, M., Pamungkas, W. J., Syuhri, S. N. H., Wibowo, R. K. K., Basuki, H. A., Djumharianto, D., Yudistiro, D., Choiron, M. A., Wardana, M. K. A., & Thongchai, S. (2025). Advanced optimization of drone frame design through the application of generative design techniques and 3D printing technology. *Journal of Mechanical Science and Technology*, *39*(1), 1–10. DOI: 10.1007/s12206-024-1210-x

García-Granero, Á., Jerí-McFarlane, S., Torres-Marí, N., Brogi, L., Ferrà-Canet, M., Navarro Zoroa, M. Á., Gamundí-Cuesta, M., & González-Argenté, F. X. (2025). 3D-reconstruction printed models and virtual reality improve teaching in oncological colorectal surgery. *Techniques in Coloproctology*, *29*(1), 24. DOI: 10.1007/s10151-024-03074-3 PMID: 39699719

Giraldo, J. P., Cho, S. S., Eghrari, N. B., Dholaria, N., Farber, S. H., Ehredt, R. B., ... & Uribe, J. S. (2025). Advances in neurosurgical education: Literature review of mixed-reality simulation models and novel mixed-reality spine prototype. Journal of Neurosurgery: Spine, 1(aop), 1-14.

Gong, Z., Chen, J., Jiao, X., Gong, H., Pan, D., Liu, L., Zhang, Y., & Tan, T. (2024). Genome-scale metabolic network models for industrial microorganisms metabolic engineering: Current advances and future prospects. *Biotechnology Advances*, *72*, 108319. DOI: 10.1016/j.biotechadv.2024.108319 PMID: 38280495

Hevko, I. V., Potapchuk, O. I., Lutsyk, I. B., Yavorska, V. V., Hiltay, L. S., & Stoliar, O. B. (2025). A novel pedagogical approach to equipping prospective IT professionals with skills in 3D modelling and reconstruction of architectural heritage. In *CEUR Workshop Proceedings* (pp. 85-101).

Isikay, I., Cekic, E., Baylarov, B., Tunc, O., & Hanalioglu, S. (2024). Narrative review of patient-specific 3D visualization and reality technologies in skull base neurosurgery: Enhancements in surgical training, planning, and navigation. *Frontiers in Surgery*, *11*, 1427844. DOI: 10.3389/fsurg.2024.1427844 PMID: 39081485

Jo, H., Lee, J. K., Lee, Y. C., & Choo, S. (2024). Generative artificial intelligence and building design: Early photorealistic render visualization of façades using local identity-trained models. *Journal of Computational Design and Engineering*, *11*(2), 85–105. DOI: 10.1093/jcde/qwae017

Kaur, M., Selvakumar, P., & Dinesh, N. Mohit, Chandel, P. S., & Manjunath, T. C. (2025). Reconstruction and preservation strategies for cultural heritage. In S. Oflazoğlu & M. Dora (Eds.), Preserving cultural heritage in post-disaster urban renewal (pp. 385-412). IGI Global Scientific Publishing.

Kommineni, M., Chundru, S., Maroju, P. K., & Selvakumar, P. (2025). Ethical implications of AI in sustainable development pedagogy. In Chinedu, C., Mafarja, N., Hussein, M., & Saleem, A. (Eds.), *Rethinking the pedagogy of sustainable development in the AI era* (pp. 17–36). IGI Global Scientific Publishing. DOI: 10.4018/979-8-3693-9062-7.ch002

Lefranc, G., Cabrera, M. P., Comparan, R. O., & López-Juárez, I. (2025). Advanced decision-making strategies and technologies for manufacturing: Case studies, and future research directions. *International Journal of Computers, Communications & Control*, *20*(1). Advance online publication. DOI: 10.15837/ijccc.2025.1.6906

Li, X., Wang, J., & Zhang, L. (2025). Gamifying data visualization in smart cities: Fostering citizen engagement in urban monitoring. *Advances in Engineering Technology Research*, *13*(1), 151–151. DOI: 10.56028/aetr.13.1.151.2025

Limbore, N. V., Anute, N., Ahluwalia, G. K., Chordiya, P. B., Kalshetti, P., & Selvakumar, P. (2025). Ethical leadership and corporate responsibility. In Harriott, S. (Ed.), *Innovative approaches to managing conflict and change in diverse work environments* (pp. 193–218). IGI Global Scientific Publishing. DOI: 10.4018/979-8-3693-9556-1.ch009

Liu, Z., Xia, F., Guo, B., Leng, C., Zhang, E., Xu, L., Chen, X., & Zhu, P. (2025). Comprehensive evaluation of the ramification patterns of hepatic vascular anatomy based on three-dimensional visualization technology. *Updates in Surgery*, *77*(8), 1–13. DOI: 10.1007/s13304-025-02064-w PMID: 39853656

Maharjan, S., Ma, C., Singh, B., Kang, H., Orive, G., Yao, J., & Zhang, Y. S. (2024). Advanced 3D imaging and organoid bioprinting for biomedical research and therapeutic applications. *Advanced Drug Delivery Reviews*, *208*, 115237. DOI: 10.1016/j.addr.2024.115237 PMID: 38447931

Maskeliūnas, R., Maqsood, S., Vaškevičius, M., & Gelšvartas, J. (2025). Fusing LiDAR and photogrammetry for accurate 3D data: A hybrid approach. *Remote Sensing (Basel)*, *17*(3), 443. DOI: 10.3390/rs17030443

Qiu, S., Xie, B., Liu, Q., & Heng, P. A. (2025, January). Advancing extended reality with 3D Gaussian splatting: Innovations and prospects. In *2025 IEEE International Conference on Artificial Intelligence and eXtended and Virtual Reality (AIxVR)* (pp. 203-208). IEEE. DOI: 10.1109/AIxVR63409.2025.00039

Sachdeva, R., Armstrong, A. K., Arnaout, R., Grosse-Wortmann, L., Han, B. K., Mertens, L., Moore, R. A., Olivieri, L. J., Parthiban, A., & Powell, A. J. (2024). Novel techniques in imaging congenital heart disease: JACC scientific statement. *Journal of the American College of Cardiology, 83*(1), 63–81. DOI: 10.1016/j.jacc.2023.10.025 PMID: 38171712

Saifi, S., & Anandakumar, R. M. (2025). Web-based visualization and rendering of aerial LiDAR point cloud for urban flood simulation. *International Journal of Disaster Resilience in the Built Environment, 16*(2), 260–274. DOI: 10.1108/IJDRBE-07-2023-0079

Samal, A., & Misra, L. I. Mohit, Pougajendy, S., Chandel, P. S., Manjunath, T. C., & Selvakumar, P. (2025). Musealization and tourism practices. In S. Oflazoğlu & M. Dora (Eds.), Preserving cultural heritage in post-disaster urban renewal (pp. 115-142). IGI Global Scientific Publishing.

Shaikh, M., & Birajdar, F. (2024). Advancements in remote sensing and GIS for sustainable groundwater monitoring: Applications, challenges, and future directions. International Journal of Research in Engineering. *Science and Management, 7*(3), 16–24.

Sharma, A., Lysenko, A., Jia, S., Boroevich, K. A., & Tsunoda, T. (2024). Advances in AI and machine learning for predictive medicine. *Journal of Human Genetics, 69*(10), 487–497. DOI: 10.1038/s10038-024-01231-y PMID: 38424184

Soundariya, A., Behera, N. K., Selvakumar, P., Chandel, P. S., Gupta, S., & Manjunath, T. C. (2025). AI and environmental ethics: The role of AI in addressing climate change ethical considerations in AI-driven environmental solutions. In Chakraborty, S. (Ed.), *Human Values, Ethics, and Dignity in the Age of Artificial Intelligence* (pp. 321–346). IGI Global Scientific Publishing. DOI: 10.4018/979-8-3693-9035-1.ch013

Sreekanthaswamy, N., Anitha, S., Singh, A., Jayadeva, S. M., Gupta, S., Manjunath, T. C., & Selvakumar, P. (2025). Digital tools and methods. In Çela, E., Paliktzoglou, V., Sengar, S., Eappen, P., & Vajjhala, N. (Eds.), *Enhancing school counseling with technology and case studies* (pp. 25–48). IGI Global Scientific Publishing. DOI: 10.4018/979-8-3693-8392-6.ch002

Tamilarasi, C., Sowmiya, K., Vignesh, G. D., Jasmine, R. L., Nehru, P., & Selvakumar, P. (2025). AI in renewable energy education: Teaching the future of sustainable power systems. In Chinedu, C., Mafarja, N., Hussein, M., & Saleem, A. (Eds.), *Rethinking the pedagogy of sustainable development in the AI era* (pp. 133–152). IGI Global Scientific Publishing. DOI: 10.4018/979-8-3693-9062-7.ch007

Uddin, M. K. S. (2024). A review of utilizing natural language processing and AI for advanced data visualization in real-time analytics. *Global Mainstream Journal*, *1*(4), 10–62304. DOI: 10.62304/ijmisds.v1i04.185

Ullah, M., Bibi, A., Wahab, A., Hamayun, S., Rehman, M. U., Khan, S. U., Awan, U. A., Riaz, N., Naeem, M., Saeed, S., & Hussain, T. (2024). Shaping the future of cardiovascular disease by 3D printing applications in stent technology and its clinical outcomes. *Current Problems in Cardiology*, *49*(1), 102039. DOI: 10.1016/j.cpcardiol.2023.102039 PMID: 37598773

Vaidyanathan, B., Maya, H. A., Xavier, S., & Kappanayil, M. (2025). 3-Dimensional virtual reality-based visualization of fetal cardiac anatomy with the use of ultrasound datasets. *JACC. Case Reports*, *30*(8), 103211. DOI: 10.1016/j.jaccas.2024.103211 PMID: 40250906

Wang, P., Rui, X., Gu, J., Huang, K., Zhou, L., & Jiang, M. (2025). Fast parametric modeling of visualized simulation and design for tracked vehicle system. *Advances in Engineering Software*, *201*, 103852. DOI: 10.1016/j.advengsoft.2024.103852

Wu, T., Yuan, Y. J., Zhang, L. X., Yang, J., Cao, Y. P., Yan, L. Q., & Gao, L. (2024). Recent advances in 3D Gaussian splatting. *Computational Visual Media*, *10*(4), 613–642. DOI: 10.1007/s41095-024-0436-y

Xu, H., Omitaomu, F., Sabri, S., Zlatanova, S., Li, X., & Song, Y. (2024). Leveraging generative AI for urban digital twins: A scoping review on the autonomous generation of urban data, scenarios, designs, and 3D city models for smart city advancement. *Urban Informatics*, *3*(1), 29. DOI: 10.1007/s44212-024-00060-w

Yiğit, A. Y., & Uysal, M. (2025). Virtual reality visualisation of automatic crack detection for bridge inspection from 3D digital twin generated by UAV photogrammetry. *Measurement*, *242*, 115931. DOI: 10.1016/j.measurement.2024.115931

Zaman, N., Ong, J., Waisberg, E., Masalkhi, M., Lee, A. G., Tavakkoli, A., & Zuckerbrod, S. (2024). Advanced visualization engineering for vision disorders: A clinically focused guide to current technology and future applications. *Annals of Biomedical Engineering*, *52*(2), 178–207. DOI: 10.1007/s10439-023-03379-8 PMID: 37861913

Zhang, L., Li, Y., Pan, Y., & Ding, L. (2024). Advanced informatic technologies for intelligent construction: A review. *Engineering Applications of Artificial Intelligence*, *137*, 109104. DOI: 10.1016/j.engappai.2024.109104

Zheng, X., Zhang, S., Lin, W., Zhang, A., Mayol-Cuevas, W. W., & Shen, J. (2025). CULTURE3D: Cultural landmarks and terrain dataset for 3D applications. arXiv preprint arXiv:2501.06927.

Zhou, Y., & Liu, J. (2024). Advances in emerging digital technologies for energy efficiency and energy integration in smart cities. *Energy and Building*, *315*, 114289. DOI: 10.1016/j.enbuild.2024.114289

Chapter 2
Digital Twin Applications in Urban Planning:
BIM and GIS Synergy for Real-Time City Simulation

R. N. Ravikumar
https://orcid.org/0009-0009-3705-1681
Marwadi University, Rajkot, India

S. Aarthi
https://orcid.org/0009-0006-9064-2091
Marwadi University, Rajkot, India

ABSTRACT

The integration of Building Information Modeling (BIM) and Geographic Information Systems (GIS) is transforming the landscape of urban planning by enabling intelligent, data-driven, and sustainable city development. This chapter explores the synergy of BIM-GIS convergence through the lens of Digital Twin applications, real-time simulation, and smart infrastructure planning. It discusses technological frameworks, standards, and visualization tools that support interoperability and spatial intelligence. Real-world case studies from Singapore and Helsinki demonstrate the practical benefits in underground utility mapping, participatory planning, and green zoning. The chapter also highlights emerging technologies such as IoT, AI, and edge computing that enhance spatial intelligence. Legal, governance, and adoption challenges are critically examined, along with future research directions. The integration of BIM and GIS offers a powerful approach to achieving resilient, efficient, and citizen-centric urban systems.

DOI: 10.4018/979-8-3373-4058-6.ch002

1. INTRODUCTION TO BIM-GIS INTEGRATION

The ability to use Building Information Modeling (BIM) and Geographic Information Systems (GIS) together is revolutionary in the context of forecasting and constructing new urban areas and expanding infrastructure. Traditional BIM is mostly used in the detailed architectural and construction planning of a structure and is particularly good at micro level detail and components such as structure and systems of a building and materials used as well as various mechanical systems. On the other hand, GIS is able to offer macro level data and spatial information which can integrate and perform terrain analysis, zoning, systems of the planning and environmental data (Andrianesi & Dimopoulou, 2020). This new technological merger offers a fully integrated digital model of fused fabricated and natural terrains which allows planners to model and analyze urban problems at a greater scale. Urban planners and other policy makers are able to utilize and manipulate this data to gain precise, structured, and logic information in order urbanize a whole area due to the synchronization of GIS and BIM technologies. This helps improve the coordination and changes that can be done in regard to land use, planning, and management of utilities, disaster management and response, as well as smart infrastructure building (Avezbaev et al., 2023). The reverse is also correct in that the integration provides a framework for the implementation of Digital Twins which are real-time hyper-realistic virtual models of urban systems. Due to the pressures of the urbanization, depletion of resources as well as the impact of climate change on the cities, BIM-GIS convergence becomes one of the key pillars of next-generation spatial intelligence that provide specific, scalable, and interoperable approaches in ensuring resilient cities in the urban future.

1.1 Understanding the Fundamentals of BIM and GIS

BIM and GIS are two not quite similar but closely related technologies. BIM is an electronic procedure which generates accurate 3D designs of the building and infrastructure and encrypts data on geometry, material, the level of construction, maintenance programs and also data on life cycle. It is majorly applied in micro level which is on design construction and management of a facility in a particular structure. On the other side GIS is a spatial technology that manipulates, processes, and visualizes geospatial information, such as terrain, transport networks, land use, hydrology, and population (Wan Abdul Basir et al., 2023a). From a macro standpoint, it assists in resolving issues of spatial ecology and mapping distant environmental relationships and their configurational parts for urban planning and policy development. For urban and regional planning augmentation, GIS offers satellite imagery, topographic maps, and census data. Appreciating Rational BIM helps comprehend the monopolistic value the stakeholders place in the Rational BIM. GIS spatial data over broader regions and accurate asset-level details in the BIM are mutually exclusive. This facilitates and reinforces model-based integrated multi-physics which, in combination with other sophisticated systems, improves spatial infrastructure and complex integrated infrastructure, Digital Twins, and smart city ecosystems.

1.2 Importance of Integration in Urban Planning and Infrastructure

The combination of BIM and GIS urban planning has integrated architecture and geomatics information into a single model autonomously. It benefited from urban planning BIM and GIS coupled real-time analysis to automate complex info cluster visualization and sim. Thus, decision-making improved for stakeholders across various scales (Wan Abdul Basir et al. 2023b). Planners are now able to assess a structure and its surroundings through a lens that considers interrelated infrastructural components and subsequent development region traffic, water, and energy system implications. Infrastructure spending results are better with the intersection of risk optimization, resource spend, and site analysis. Further, the integrated use of BIM and GIS successfully merges with the real-time twin technology to create a compute defined virtual representation of systems to design with. These systems support the planning and design of adaptive infrastructure for reactivation along with maintenance and fire emergency strategies. BIM-GIS Integration will be essential for developing a city. which is capable of coping well with urbanization and climate change as well as capable of facing future pressures.

1.3 Background Study

Most fast-moving city population predicts over 70% shift in people moving to city by 2050. This is putting unsustainable pressure on the infrastructure, environment, and governance of cities. Here, the use of planning, integrated thinking, and the use of technology is lifesaving. The use of BIM and GIS together for these urban problems constructs an integral system for real-time analysis, planning response, sustainable development and integration for dynamic urban systems (I. H. Hijazi et al., 2020). The use of BIM and GIS has been used privately. BIM occupies itself with complex structures and infrastructure, planning the geometry, construction phases, and operational phases of the object. In turn, GIS is concerned only with the management and analysis of spatial data. and used in conjunction with other construction systems.

GIS technology supports 'green' investigations, as well as transport planning, zoning, and demographic studies, in graphical representation and pattern analysis in relation to broad regional frameworks. GIS technology enables planners to study the underlying and surface infrastructure, land use, flood-prone areas, the transportation network, and population concentration areas (I. Hijazi et al., 2010). The possibility of corresponding integrated BIM and GIS technologies reinforces these gains to create a well-balanced digital ecosystem not just on spatial reasoning but also on rich semantics. The stakeholders can now, through this synergy, visualize and forecast the behavior of buildings in relation to their environment and spatial surroundings, and also process and analyze such relations to improve their decision making in any stage of urban development. These integrated models can be used in a live responsive manner when the models are extended system using Digital Twin technology to reflect the current real-world state in real world terms with data taken by IoT sensors, remotely sensed, and cloud-based analytics used to drive the model.

Traditionally, the drawbacks of the BIM-GIS integration have been data structure, the poorly harmonized coordinate systems and interoperability standards (Hor et al., 2016). BIM relies on local coordinate systems and works in detail on high resolution, whereas GIS relies on global reference and work on large scale geographical datasets. Nevertheless, improved open data standards like IFC (Industry Foundation Classes) and CityGML have directly enhanced interaction between the two systems, greatly. Also, the development of software systems like Autodesk infraWorks, Esri ArcGIS GeoBIM, and FME are making it more and more practical to close the gap between BIM and GIS in practical and scalable applications (Kang, 2023). The increase in attention towards smart cities, climate adaptation, and resilient infrastructure has inflated the rapid rise in the convergence of BIM and GIS. This interaction from different sectors on the Urban solar Potential Planners fosters the use in flood simulation, traffic engineering, and in the compliance of the construction and operation of the works.

The use of 3D immersive environments with the stakeholders allows the planning of the development in advance which enables them to design the master plan. This chapter focuses on the increasing need to streamline the urban planning process as a result of rapid urbanization alongside increasing environmental concerns. The integration of Real time technology and BIM GIS espouses the construction of the new smart infrastructures that are inclusive and sustainable (Kiavarz et al, 2018). The union of these technologies is of a strategic nature besides being of new nature, as it transcends across the spatial domains of governance, urban design, and the participation of citizens in the future disposition of the city. The key drivers of BIM GIS integration with Rapid Urban Growth are illustrated in Figure 1. It also depicts the application of digital technologies in the planning for sustainability of the engagement and rationalization of the infrastructure.

Figure 1. Impact of BIM-GIS Synergy and Urban Growth Trends

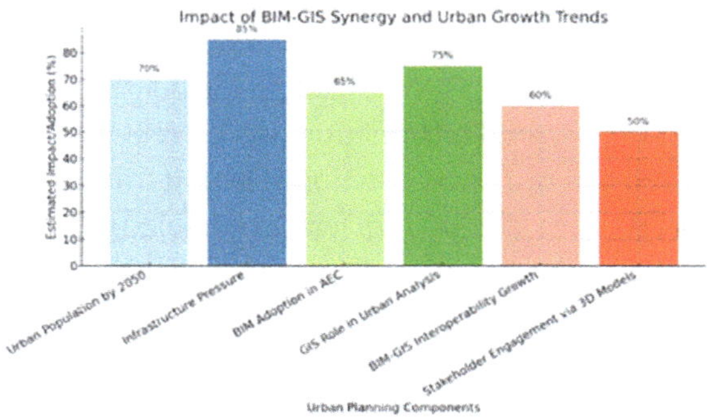

1.4 Literature Review

The adoption of unified smart city development and infrastructure maintenance has posed complex problems with interface design, database management, and information fusion of Building Information Modeling with Geographic Information Systems. More and more articles suggest the optimistic perspective associated with the merger of BIM and GIS technology with regard to the advancement of urban ecosystems. In this section, a review of the most significant studies and contributions are made on the technological, methodological, and practical basis of this convergence. The researchers of the early literature stressed the necessity of interoperability between the BIM and GIS and the semantic and geometric differences between them as a problem that hindered the exchange of information. Their study defined the incompatibilities in data structure as BIM concentrated on object-based modeling, high precision at a local coordinate system whereas GIS worked on spatial relationships at the global level. To a certain degree, some of these fears were overcome by the introduction of open standards to BIM (International Alliance for Interoperability, IFC) and GIS (CityGML). (Boccardo et al., 2024) investigated the CityGML applications in producing semantically rich 3D city models and how these models could be aligned with the BIM datasets that would facilitate urban simulations. On the same note, a study by (Liu et al., 2017) revealed that the combination of BIM and GIS enhances the process of managing the life cycle of infrastructures by having the relative location of an asset based on details. Their case study of construction projects involving subways demonstrated a high level of visualization of subway projects, a better communication with the stakeholders, and minimization of errors caused by spatial inconsistencies.

The past years have seen a paradigm shift in the literature on BIM-GIS towards application on digital twins and smart cities (Narindri et al., 2022) investigated the technologies to adapt BIM and GIS to be used in the construction logistic management in cities, as real-time spatial information was helpful in the coordination of the construction schedules and traffic congestion. The latter is extended to the Digital Twin technology by (Batty et al., 2012) in terms of tracking complex urban systems through real-time monitoring and modeling by connecting these systems with the urban data platforms. These studies point out that BIM-GIS integration is not only used during design and construction, but also during continuing operations, asset maintenance and emergencies. The IoT analytics of devices in real time in real world applications with Digital Twins in BIM GIS tracked energy modeling of projects, mobility, and air quality improvement (Zhao et al., 2019b) discussed the BIM-GIS based Digital Twins and the IoT devices real time sensor data.

The writings Predictive Digital Twinning and the Automation of Cities emphasized the possible functions of Machine Learning and AI algorithms in zoning automation, predicting urban area infrastructure failures, and digitally planning urban spaces in real time. Sun and (Sladić et al., 2021) showcased the 3D GIS BIM models, VR, and Web GIS for citizen participatory planning as an urban planner. All these enhance decision-making to be more open and participatory. However, the literature underlines the issues of data governance, the legal boundaries, and cross system integration of the Digital Twin implementations of megapolises. There are no doubts left about the BIM GIS seamless integration as it is a prerequisite for the next generation of spatial intelligence. Based on the research available, the shift from fragmented models to real-time, holistic, citizen-centric frameworks due to planning data and digital technology innovations is paradigm-altering.

2. TECHNOLOGICAL FRAMEWORKS FOR BIM-GIS CONVERGENCE

The connection between BIM and GIS rests on sophisticated technology capable of interfacing, sharing, and collaborating on data in real-time. This includes data model specifications, custom interfacing works, and sufficiently large computing resources needed for augmented shift from fragmented models to intelligent integrated networks (Leng & Zhen-Zhong, 2020). Major aids to this are Industry Foundation Classes (IFC) as developed as BIM (and the well-known standard CityGML as a GIS standard) which standardize data representations and facilitated semantic interoperability (Liu et al., 2017). The process is also accelerated by cloud computing,

IoT networks and 5G communication that enables high-speed data-transmission, real-time update and distributed-processing. Digital Twins also can rely on edge computing and containerized microservices that make the information responsive by performing local tasks near the data providers. Other technological systems consist of BIM-GIS connected tools like Autodesk Forge, Esri ArcGIS, and open-source tools, including, but not limited to, FME or GeoServer. Immersive, real-time simulations are made possible by the use of visualization tools on a game engine (e.g., Unity or Unreal Engine). Well put simply, BIM-GIS technological architecture is the skeleton of smart city modelling-the concept that will turn the city planning process into paperless modelling incorporating an ecosystem of data-driven planning that favors long-term sustainability, teamwork, and forecasting.

2.1 Software Tools and Data Models for Interoperability

The interface between BIM and GIS will only be seamless when there are software tools to translate and synchronize data in various environments. These bridges promote interoperability whereby the models of the objects that BIM uses in its brood storage and GIS geodata functions can equally work in a single digital environment. Among them are ArcGIS GeoBIM by Esri, Autodesk Revit, InfraWorks and FME (Feature Manipulation Engine). ArcGIS GeoBIM is the technology that allows the stakeholders to read and explore BIM data in a geographic context (connecting with GIS based spatial information) (Maltseva et al., 2024). Autodesk InfraWorks is used to visualize every node of the big-scale infrastructure projects through linking the Revit models to geographic data. FME has flexible data transformation and integration workflows across a number of data formats and has proprietary and open-source data model support. Data models are highly important in consistency and context. In the case of BIM, a standard representation of building components is possible using the schema of the IFC (Industry Foundation Classes). In one side, CityGML describes urban objects geometry-accurately and semantically in GIS. Such data models guarantee such similar platform interpretations of features roads, utilities, and structures will permit consistent planning, analysis, and simulation processes (Song et al., 2017). The presented Table 1 indicates the most frequently utilized tools and data models ensuring the interoperability between BIM and GIS. It details how their types differ, their roles and functions in achieving smooth integration to facilitate effective, collaborative and smart urban planning.

Table 1. Key Software Tools and Data Models for BIM-GIS Interoperability

Tool / Model	Type	Function	Integration Role
Autodesk InfraWorks	Software	Creates infrastructure design models and connects them with GIS data	Bridges BIM models with geospatial context for urban infrastructure planning
Esri ArcGIS GeoBIM	Software	Links BIM models (e.g., Revit) with GIS layers in ArcGIS environment	Enables spatial analysis and visualization of BIM data in real-world settings
FME (Feature Manipulation Engine)	Middleware/ETL Tool	Transforms and integrates spatial and non-spatial data formats	Converts between GIS and BIM formats like IFC and CityGML for interoperability
IFC (Industry Foundation Classes)	Data Model	Open standard for representing building and infrastructure data	Enables exchange of BIM data across platforms and disciplines
CityGML	Data Model	Open format for 3D city modeling with semantic and spatial components	Facilitates detailed representation of urban environments in GIS systems

2.2 Standards and Protocols (e.g., IFC, CityGML) Enabling Integration

BIM-GIS integration incorporates standardization as the foundation. In the absence of shared protocols, transferring data between BIM and GIS platforms would be ineffective and inefficient. Among the most frequent standards which make such integration possible, one can mention IFC (Industry Foundation Classes) and CityGML (City Geography Markup Language). BuildingSMART created IFC to be an open, unbiased data format that allows sharing of the building infra structure data among various software applications. It offers architectural, structural and service systems in detail descriptions which are important in BIM applications (Tan et al., 2023). It facilitates the compatibility of tools like Revit, ArchiCAD, Tekla, and Navisworks which means that construction and building specifics are kept in the same line over platforms. CityGML, which has been designed by the Open Geospatial Consortium (OGC), is a sematic 3D model that is specifically used to represent the urban setting. It allows it to model buildings, terrain, vegetation, and transport systems and their attribute data; therefore, it is very applicable in a GIS environment (Tobiáš, 2016). Middle ware and APIs are growing support of these standards, and thus data flows become consistent across domains. With Digital Twins in mind, the convergence of IFC and CityGML is the key to precise, scale, and time generated physical asset representations in the real-world landscape.

3. 3D URBAN MODELING AND VISUALIZATION

Visual and analytical correlation of BIM-GIS convergence is 3D urban modeling and visualization, which allows planners, architects, and policymakers, to get a better perception of the urban development environment with respect to space, structure and environment. The integration of constructional details depicted in Building Information Modeling through rich 3D models with geometric information system spanning blanket GIS coverage enables the construction of lifelike visualizations at the scale of entire districts, cities, or even entire infrastructures. The dynamic interactive environments that 3D models with depth augment include real time information IoT devices, simulations, and predictive analytics (Zhu & Wu, 2021). Integration of multiple approaches of integration fosters collaborative work and improves integration with decision making interfaces pertinent to them. Decision making models can be focused with specific congestion vehicle and pedestrian simulations, 3D shadowing of flood risk during different times of the year, and more. These exercises are more manageable with 3D models. Interfaces to the Internet enable real time dynamic smart city environments to be accessed through Advanced game technology (Unity, Unreal Engine), WebGL, 3D GIS (CesiumJS, Esri CityEngine), and others (Zhu et al., 2021). 3D models have value in augmentation of participation and societal engagement. With VR devices and mobile phones, public can have more effective engagement to access and understand the proposed changes. This, in turn, increases the trust of the people since they can visualize the planned and proposed changes more effectively. 3D urban modeling serves as a bridge between technical planning and public understanding and turns data into narratives that give shape to cities that are more resilient and more intelligent.

3.1 Generation of 3D City Models Using BIM-GIS Synergy

The synergy between BIM and GIS arises at the production of 3D city models which adds the architectural detail, spatial elements and geospatial extent to the creation of accurate city representations in multiple dimensions. BIM provides accurate 3D models and semantic information about specific site building, infrastructure-cabinets, walls, doors, etc. and GIS provides the topography, zoning of land, and transport systems, terrain features (Borkowski et al., 2022). This helps automate construction of metropolitan scale models as instruments like Esri CityEngine, Autodesk InfraWorks, and FME consolidate numerous data streams into cohesive digital ecosystems. Multilayered analytical models, domain specific digital twins, possess depth beyond 3D visuals and aid in simulation, analysis and real time alterations. Enhanced interactive data visualization augments model interactivity. These models are also compatible with CityGML and IFC formats, as well as other open data standards. Additionally, real-time streams and cross-referenced Time-Space-Information Models enable predictive regulation of IoT systems for utilities, energy consumption, and public safety (Cao et al., 2023). Thus, the synergies of BIM with GIS allow for comprehensive digital twins of cities with holistic spatial, sematic, and dynamic geo-referenced, real-world counterparts for all construction elements. Figure 2 demonstrates the role of GIS and BIM integration in developing 3D models of cities. It shows the synergistic advantages of GIS (macroscopic spatial information) and BIM (microscopic details of structures) in creating dynamic, information-laden urban systems with real-time IoT access.

Figure 2. Conceptual Framework of BIM-GIS Synergy for 3D City Modeling

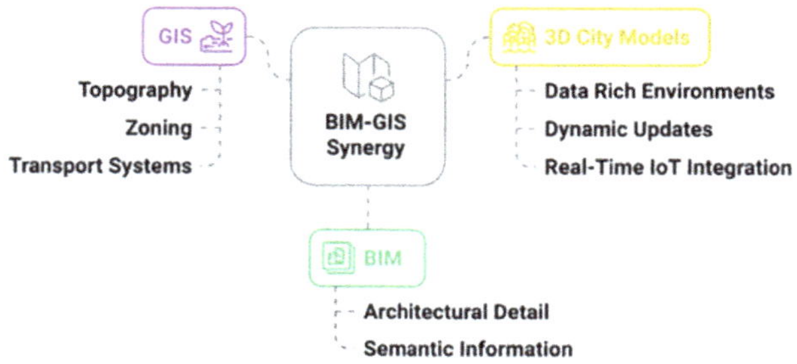

3.2 Immersive Visualization for Stakeholder Engagement

Visualizing immersive environments enhances stakeholder participation in urban planning. Engaging interactive environments translate complex urban big data into user-friendly models for stakeholders such as policymakers, engineers, investors, and citizens, allowing them to better understand and appreciate contributions accessed freely to the suggested developments. After the models are built, automated texts, lighting, and interactivity can make cross-platform 3D models from BIM-GIS realistically immersive using Unity, Unreal Engine, CesiumJS or similar. Users can traverse scenarios in which particular infrastructures are constructed or freely navigate 3D models scaffolded on the developments using VR, AR, or web-based interfaces. They can also visualize developments from various perspectives as suggested by (Ding et al., 2020). Such hyper-realistic experiences significantly increases public participation by demystifying and simplifying the complex planning processes. They can see novel road design, landscapes or flood prevention systems and give a well-informed response. In the eyes of the decision-makers, the scenario analysis can be enhanced by incorporating the immersive models to allow multiple decisions to be reviewed in a simulated environment. Immersive visualization allows transparency, trust, and collaborative planning with lay stakeholders since the communication barrier between technical experts and lay stakeholders is closed. It opens up data and turns it into common experiences, such as that the urban development is smart and efficient, inclusive, and participatory.

4. SUSTAINABLE PLANNING AND ENVIRONMENTAL IMPACT

Combining BIM and GIS is a ground-breaking tool toward better sustainable urban planning as it accurately analyses, simulates, and optimizes environmental effects of a city. BIM can deliver micro-scale data in terms of building contents, energy usage and whole life costs whereas GIS can deliver Macro scale i.e. facts on the land use, terrain, hydrological features and climatic conditions (El Mekawy et al., 2021). Coupled together, the synergy enables the evaluations of both individual dwellings and urbanizing infrastructures to determine their ecological footprints. With this combined model, the urban planners will be able to model the sun exposure as well as the heat islands and air quality dispersion, storm water runoff and green infrastructural locations. BIM-GIS platforms enable the creation of scenario theory amid the drop in carbon emissions and the empowerment of transportation modes and the improvement of biodiversity in urban environments. An example is the planners can compare the various zoning patterns or material to be used in the construction process to be able to see their environmental trade-offs before applying them (Meschini et al., 2022). Digital twins of BIM - GIS models permit continuous surveillance of a sector and advanced planning. Model Adjustments in real-time by adding IoT sensor data like energy mining and air quality sensors data, can quicken rectifications. Thus, the BIM GIS integration meets several of the sustainable development goals. It optimizes the urban system, strengthens, adapts to climate challenges and supports urban areas to transition to a low-carbon and sustainable development.

4.1 Using BIM-GIS for Environmental Performance Simulation

Another key use case in urban design sustainability is evaluating projected sustainability outcomes, which can be aided in part by the combined use of BIM and GIS. BIM uniquely contains reliable geometric and material characteristics of buildings which can be used in simulated energy models, daylight modeling, and life-cycle assessments. GIS geostatic data sets which consist of solar radiation, wind geospatial data, and elevation geospatial data, allow urban designers to model and assess potential environmental conditions with greater simulation fidelity and contextual situational awareness (Sani et al., 2022). Tools similar to Assessment of design alternatives with Environmental indicators include Autodesk Insight, Revit Energy Analysis, and ArcGIS Urban. As an example, the influence of the orientation of buildings on solar heat gain or the influence of local topography on wind flow and ventilation can be simulated by a planner. Passive design strategies which are directed at minimizing the use of mechanical systems and encouraging building envelope energy efficiency rely on this simulation. Furthermore, the synergy facilitates the compliance with regulatory checks in green buildings like LEED, BREEAM and IGBC due to its data-driven intelligence in regard to the thermal performance, carbon footprint and resource utilization (Tan et al., 2023). BIM-GIS environmental simulation tools assist designers and policymakers to engineer low-carbon resource-efficient urban environments, which is possible because of by virtue of multi-scale analysis features: they can be used to analyze single structures, but the whole district as well.

4.2 Green Infrastructure and Smart Energy Systems Mapping

Cities planning to achieve targets of green and futuristic imperatives need to map and manage green infrastructure and smart energy systems. Urban planning BIM-GIS integration enables urban planners to geolocate the position, performance, and connectivity of the green infrastructure assets green roofs, urban forests, bioswales, and wetlands, as well as utilities such as power grid, photovoltaics, and district heating grids. Using GIS layers, the macro-scale environmental information and spatial analysis can be accessed including the hydrological flows and the suitability of the land to green interventions (Usmani et al., 2021). BIM adds micro level data in terms of material properties, structural support and space geometry that allow planners to plan structural viable and environmentally friendly green infrastructure elements. Basically, in the context of energy systems, such an integration can be used to model renewable energy system performance and geographical location (solar panels, wind turbines), electric vehicle charging stations infrastructure, and smart grid. The model is also further augmented with real-time data provided by smart meters and IoT sensors to allow dynamic load balancing, real-time maintenance (Vacca et al., 2018). Such holistic mapping solution can help discover urban heat islands, maximize the energy distribution, and facilitate the usage of nature-based solutions to make cities stronger. As the cities are switched to net-zero pathways, BIM-GIS-enabled models will be used to guide decisions on the implementation of integrated green and energy infrastructure networks. The figure 3 outlines the stepwise progression from inefficient urban planning to optimized cities using geo-location mapping, detailed data integration, and real-time updates. Mapping green infrastructure while considering energy systems helps in the creation of resilient and dynamic urban ecosystems and facilitates more streamlined operations.

Figure 3. Green Infrastructure and Smart Energy Mapping for Urban Optimization

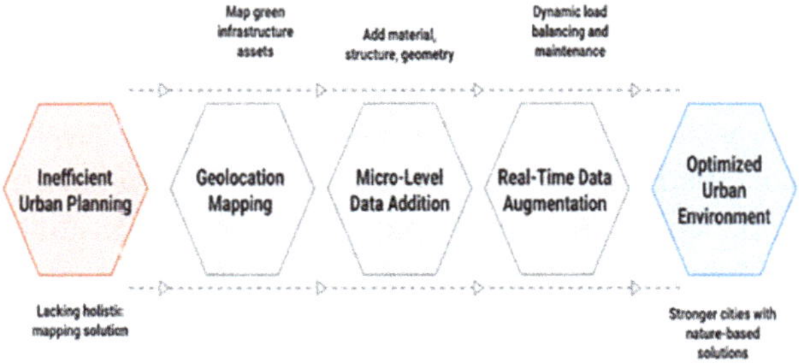

5. INFRASTRUCTURE ASSET MANAGEMENT AND OPERATIONS

The construction will be held on the backbone of the civil engineering regarding asset management and operational infrastructures which will be set supporting the design and management processes of the systems. Vertical (building, bridge) and horizontal (road, utility) assets of the infrastructure will be managed seamlessly. Panning over the city, BIM GIS in urban dry in planning offers intricate geometry with the asset repair history. GIS specializes in the integration of space geometry with urban networks (Zhao et al., 2019a). Give integrated functions advanced ability to forecast predictive maintained with precision timing, and to optimize the integrated in the plant maintenance and operations functions. For example, GIS describes Water Utilities with graphical representation and sensor inputs to BIM models indicates pipes at risk of leakage. Also, facility managers monitor HVAC and artificial lighting with IoT BIM models and GIS positions these buildings within the environmental and zoning regulations. Furthermore, asset operations benefit from the resolution of live sensor streams to history and simulations in the digital twin environment (Zhu et al., 2018). These digital ecosystems are also used to identify anomalies on the system, to send repair crews, and assist with energy and resource savings strategies. Integrating GIS and BIM transforms the management of urban assets by allowing to add spatial intelligence to the information about the assets and make it more detailed, which makes asset management more modern and cost-effective, increasing the life cycle of the city infrastructure and providing continuity of life support in all areas of the city.

5.1 Monitoring and Maintenance of Urban Utilities

The lifelines of a modern city are its urban utilities e.g. water supply networks, sewerage systems, electricity distribution systems and telecommunication infrastructure. Efficiency of monitoring and maintaining these assets is achieved by having access to detailed engineering data, as well as having geospatial context of these assets, both of which are exactly provided when integrating BIM with GIS (Onungwa et al., 2021). Physical structure of utility systems, components into which it is broken down and service records are reflected in BIM models whereas the spatial arrangement of the utility structure, its interconnections and overlays such as environmental data are as represented in GIS layers. Collaborative Ventures are going to give the possibility for the utility operators to tackle real world maps to detailed complex engineering diagrams to view subsurface pipelines, transformer stations, and drainages. With real-time monitors for water pressure data, IOT devices, belonging to the engineering of broad-gauge flow monitoring and gas leakage, help devices with auto alerts for fault detection. This enables the devices to operate base condition and timely maintenance (Colucci et al., 2023). Furthermore, with the aid of predictive analytics, predicting equipment malfunction or system chokepoints using AI/ML based on the usage and performance historical data becomes possible. It guarantees the service interruptions are minimized, the operational expenses are lowered, and the lifespan of the assets is maximized. Supporting the monitoring performance of BIM-GIS enables the fulfilling of the urban control, safety and sustainable standards which allow the cities to function optimally and respond to the increasing demands.

5.2 Real-Time Data Integration for Smart City Applications

Integrating infrastructural systems with real-time information capture in smart cities is vital for city management effectiveness. The virtual spine where BIM, GIS Convergence sustains integration of other real-time data feeds, IoT data streams, surveillance drones, meteorological stations, and traffic data into useful spatial-temporal frameworks is amazing. The BIM designed and built construction and infrastructure models is linked with GIS urban layers which empowers city managers to control, from the metaverse, city wide traffic and energy management, ditch, waste and emergency services, and other inner-city services from a digital twin centralized dashboard. Multi-layered city models with real-time dynamic alerts about pending critical events, and spatial-temporal data synthesis enable real-time response to simulated traffic, energy, and sustainment floods (superimposed from available datasets) and other spatial-temporal critical events peak energy demands. Systems to integrate, process, and display such live streams of data include Esri ArcGIS GeoEvent Server, Autodesk Tandem, and Azure Digital Twins. The composed supplement of artificial intelligence does expand the situational awareness, to apply predictive and away-off holding of city systems (Fang et al., 2024). Lastly, it is recognized that real-time integration of data through BIM-GIS renders the smart cities responsive rather than reactive, thus enhancing the safety of the population, sustainability, and the delivery of services, even in an increasingly sophisticated urban environment.

6. LEGAL, REGULATORY, AND GOVERNANCE ASPECTS

As the importance of BIM and GIS for developing Digital Twins and planning smart cities grows, a more complex, perhaps even chaotic, legal, institutional, and governance framework is developing. All these aspects matter in relation to the governance framework, in particular the management of the bounds of the use of sensitive information, digital infrastructure, and smart systems. City information, especially in combination with real-time sensors and public and private infrastructures, is to be pampered with a great deal of privacy, ownership, and security (Zhu et al., 2020). Data governance is a major issue where governing entities such as its custodianship, access rights and data-sharing rules between governmental agencies, individual developers and citizens are to be established. To build trust in the population, the most important strategy is regulatory compliance with such frameworks as GDPR (General Data Protection Regulation), Digital Personal Data Protection Act in India, and ISO standards related to BIM (ISO 19650) and geospatial data management. Urban authorities should also come up with policies, which ensure that their systems are interoperable, systems are required to utilize open standards and accountability of AI powered simulation or decisions made by the using Digital Twins (Varlk & Dursun, 2024). The equity and transparency aspects of governance models should also be brought up by ensuring that everyone in the nation receives the benefits of spatial intelligence technologies but is not banned due to some form of digital discrimination and/or excessive surveillance. The systematization and clause within the frameworks of establishment such BIM-GIS occupied smart city regions bears considerable mark on the approbation and the past triumph of the same.

6.1 Land Use Planning and Zoning Compliance

BIM-GIS has great value in land use planning regulation compliance, which is one of the subdivisions of urban management. For example, in the case of BIM, regional 3D representations of structures' volumetric designs, altitude, function, and construction materials, sculptures integrated within the topography, and trees, are capable of streamlining construction. On the other hand, GIS integrates zoning, construction, heritage, and eco-buffering zones into a single regional grid. Under this synergy, the decision-makers can determine whether subordinated delineated new structures can meet the zoning standards such as Floor Area Ratio (FAR) and buffered building limit and height. Real-time rules-based engines autonomously perform complex evaluative actions BIM and GIS zoning as built layers and legal boundaries & rules databases. Having such capabilities helps in easily granting building designing permits, lowers legal conflicts, and helps urban expansion within a short period of time. In addition, the fusion of BIM-GIS with land tenure and land registration systems significantly raises the transparency of urban planning. It helps in compliance checking, encroachment detection, and examination of land useful for redevelopment as well as environmental impact assessment. Cities are able to move from reactive planning to proactive planning as a result of the application of integrated regulatory enforcement which is digitally modeled and can be simulated, validated, and analyzed spatially (Dursun et al., 2022).

6.2 Digital Twin Applications in Policy and Decision Making

The combination of BIM and GIS is transforming decision making in urban policy and governance in real-time by using data to make definitive choices through the application of Digital Twin technology. Digital Twins represent dynamic, iterative clones of real-life cities that support the visualization of the effects of changes in policy once enacted. One common example is the policy maker can create a Digital Twin and evaluate the consequences of traffic access, zoning, storage facility modernization, or even environmental policies (El Mekawy et al., 2021). With this predictive modeling they are able to assess various policy options bearing quantifiable consequences e.g. decrease in emissions, traffic congestions or energy use. Such models can aid in visualizing the socio-economics of urban renewal projects to ensure the fulfillment of equity and inclusivity objectives. Citywide dashboards that collate real time data of various disparate city systems such as transport, utility as well as public safety systems, and the rest of the systems help agencies poise and hence, make better, more informed, cross functional decisions targeted toward better efficiency. The public can engage in active citizenship and governance though open Digital Twins wherein users can browse policy proposals, add comments, and thereby facilitate greater transparency and engagement in governance. More strategically, the Digital Twin technologies facilitates the shift from policy planning approaches which are strategically rooted in static documents to ones that are dynamic and responsive to change. Urban decision makers are then able to compromise on the planning and incorporate greater strategy and transparency that are in alignment with long term policies on sustainability, resilience, and urban smart growth.

7. FUTURE DIRECTIONS AND RESEARCH CHALLENGES

The autonomous development of Digital Twins Interface BIM-GIS Integration turns the focus of research onto the use of real time data fusion, machine learning, and AI. Further research works are required to solve the issues of cross platform semantic interoperability, standardization, and the extension of the research to mega cities. Data privacy and cyber security along with ethical governance of the integrated systems remains primary (Arroyo Ohori et al., 2018). New fields such as quantum computing, edge-AI, and decentralized data ownership (e. g., blockchain) are promising. As well, the modeling of urban systems as an inclusive process combining social, behavioural, and economic aspects applied to Digital Twins will be critical. The coming era of smart urban research will include bridging the technical innovation with policy, accessibility and resilience.

7.1 Barriers to Large-Scale Adoption of BIM-GIS Integration

Even though the massive implementation of BIM-GIS integration has the potential to change the face of construction, there are a few technical, organizational, and regulatory hiccups on the way. Data interoperability is one of the key problems interfering with the BIM and GIS systems, as they are commonly unable to share and align the data using various formats and semantics and data hierarchy (Onungwa et al., 2021). There are also outdated systems and isolated digital business processes which are obstacles to the adoption of unified platforms in use by the city departments and external parties. Acquisition costs of the technology, differences in skills among the planners and engineers who work in the urban centres and the unwillingness to change also constrain adoption. Furthermore, the absence of robust, publicly articulated frameworks on the governance of privacy, the management of data, and the circulation of information among government, agency, and commercial actors compounds the enduring deficits of accountability and security. A problem that comes up is infrastructure's capacity to continuously scale under the rapid city expansion and evolution megacities experience. These obstacles cannot be specified to only technological improvement; they require collaboration across different fields, the embrace of standard framework archetypes, and the policy reorientation to systemic interoperability and equal access to both design and foundational policy principles.

7.2 Emerging Technologies: IoT, AI, and Edge Computing in Spatial Intelligence

These technologies such as the Internet of Things (IoT), Artificial Intelligence (AI), and edge computing have completely changed the frameworks through which we engage with spatial intelligence. IoT devices such as the air quality monitors, smart utility meters and traffic cameras, are continuously feeding information into our BIM-GIS mashup and enriching it into real-time, interactive models to assist city planners in real time. These devices can monitor and manage assets, dynamically control traffic, as well as provide warning notifications of impending power outages, hazardous air quality and alarms due to a failing electrical grid. AI and machine learning are able to investigate mountains of data and catch patterns within the data to present the critical information we need to facilitate our decision-making process. These devices can also perform Digital Twin automation relating to zoning compliance checks, their energy use, and assessing their impact on the climate. Edge computing advances this by shortening the time needed to process which is sent to cloud servers, critical in time sensitive situations, like disaster recovery. All of these, paired with other IoT devices, help construct the hyper connected, intelligent ecosystems of the future, whiles these devices offer interactive operational models of space which is dynamically changing. This restructured architecture is termed as spatial intelligence (P. Haveman et al., 2024). For capsule summary of these technologies in relation to automated actions, refer to distinct actions of IoT, Edge Cloud, and AI in relation to their role, use, and BIM-GIS Digital Twin ecosystems in Smart cities in Table 2.

Table 2. Emerging Technologies Enhancing Spatial Intelligence in Urban Planning

Technology	Functionality	Urban Application Example	Impact on BIM-GIS Integration
IoT (Internet of Things)	Real-time data collection from sensors embedded in infrastructure and environment	Monitoring air quality, traffic flow, energy consumption	Enhances live updates in Digital Twins for real-time simulation
AI (Artificial Intelligence)	Data analysis, pattern recognition, and predictive modeling	Forecasting infrastructure wear, optimizing zoning layouts	Enables automation, decision support, and smart scenario simulations
Edge Computing	Decentralized data processing close to data sources	Instant response for smart traffic lights or emergency systems	Reduces latency and supports real-time applications in city operations

8. CASE STUDY SECTION

The practical applications for integrating BIM and GIS in urban settings is highlighted through the two pioneering case studies of Singapore and Helsinki which espouse the urban environment sustainability through the older BIM GIS integration at Spatial Intelligence and the 3D Visualization Interaction Interfacing Framework. The integration aided in the Marina Bay Singapore's underground utility mapping and the sustainable redevelopment where the geospatial BIM attributed provided effective underground infrastructure integration and construction risk mitigation and associated sustainable construction strategies. At the same time Helsinki urbanised employing the BIM GIS integration for regeneration city models incorporating modular participatory planning frameworks and enforcing green zoning planning governance and fostering a multimodal mobility network. The two cities also exemplify the effectiveness of Digital Twins in collaboration, stakeholder outreach, integration, resiliency, and greener collaboration of developments. In addition to this, these case studies will not only highlight the feasible advantages of the combined digital planning but it could also act as models to communities all over the world to deploy similar types of urban development plans that are future-proof and data-driven (Sladić et al., 2021).

8.1 Case Study 1: Smart Infrastructure Development in Singapore

Singapore is now a model of smart infrastructure development across the globe, and Marina Bay community is one of the best instances of proper utilization of integration of BIM-GIS. Within the context of the Smart Nation initiative the Urban Redevelopment Authority (URA) of Singapore worked with the Singapore Land Authority (SLA) and other partners to develop a centralised online platform that combines the power of Building Information Modeling (BIM) and Geographic Information Systems (GIS) to achieve urban planning and optimise infrastructure. Underground utility mapping is one of the most important applications where the 3D BIM-GIS models were used to accurately map and digitize more than 85% of the existing infrastructure underground such as electrical lines, water mains and fibre optic cables. This limited excavation-caused conflicts considerably and coordination of utility agencies improved and reduced the delays of projects by 30% and 20% on the cost of rework.

In sustainable urban redevelopment, the integration of the BIM-GIS was to facilitate such simulation capabilities as solar exposure, wind path, and green coverage resulting in the optimizing of the building orientation and facade design. This helped to increase energy efficiency within new developments structures within

Marina Bay by 15%. In this case, digital efficiency significantly contributed to the increased allocation of 12% of the space to both pedestrian area and green zones by the planners, thus, ensuring the improvement of an urban environment liveability and improving emissions of cars (Usmani et al., 2021). The Marina Bay Digital Twin also utilized IoT data streams and was aligned with 3D National Mapping system of Singapore so the real-time construction activity and health of infrastructure could be tracked. Such innovations were used to facilitate more resilient and sustainable data-driven planning process. To conclude, Marina Bay in Singapore proves that complex digital processes involving BIM (and GIS) application can be used to de-risk infrastructure investments, enhance environmental performance, and support mutually transparent cross-agency cooperation on local infrastructure development on a global scale.

The Figure 4 shows the quantifiable benefits as a result of incorporating Building Information Modeling (BIM) and Geographic Information Systems (GIS) in the building of a smart infrastructure in Marina Bay, Singapore. Significant performance areas like mapping of underground utilities, cost of issues related to delays in project, rework costs, energy efficiency and improvement of urban liability are measured in terms of % wage increase. As the data indicates, the efficiency has increased massively, with 85% digital mapping of the subsurface utilities and 30% decrease in delays in projects, illustrating the power of BIM-GIS synergy to promote planning precision, sustainability and coordination of operations in the development of smart cities.

Figure 4. Impact of BIM-GIS Integration in Singapore's Marina Bay Development

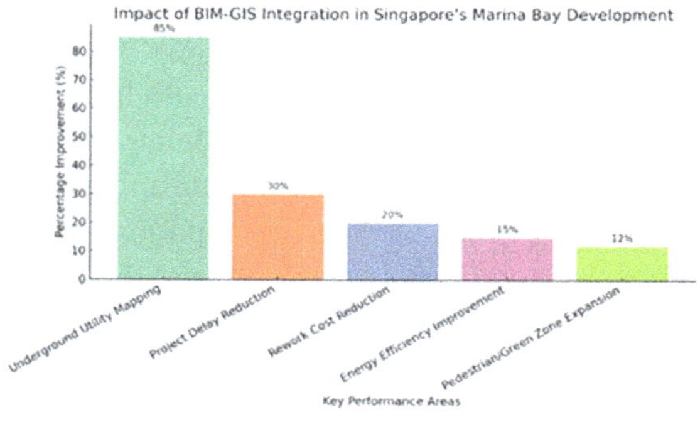

8.2 Case Study 2: Urban Regeneration in Helsinki

Finland The capital of Finland called crime Helsinki is a renowned place in the world of urban regeneration, which is done innovatively through use of 3D BIM-GIS models. The transparency, sustainability, and participation of the citizens enjoyed by the city are under the Prepared in Helsinki 3D+ project which integrates detailed Building Information Modeling (BIM) and large-scale Geographic Information Systems (GIS) to facilitate participatory urban planning and smart growth. The green zoning and the optimization of land use could be pointed out among the primary areas of interest where the BIM-GIS instruments allowed planners to model land cover, evaluate the effects on the environment, and distribute resources most efficiently. Consequently, regenerated areas showed an 18% boost in green spaces alongside with 25% of the redevelopment spaces being selected as parks, woodland towns and routes of biodiversity (Song et al., 2017). Such zones are used in stormwater where they play a mitigative role and in the urban microclimates.

Mobility was also optimized in the city. Helsinki planners combined real-time information about transportation patterns in the city with spatial models of its buildings to map pedestrian bottlenecks, cycling demand hotspots and weaknesses in the public transport network. It led to 22% decrease in the average commute time and 30% growth in bicycle infrastructure in pilot districts. Participatory planning was one of the most influential parts of the Helsinki strategy of urban regeneration. The community members were welcome to navigate planned improvements in a 3D model of the community and leave feedback on the platform. More than 60% of the local citizens in the feed areas have been involved in the feedback process which has affected the zoning, design and accessibility facets (Wang, 2021). This encompassing strategy improved the trust and congruency between the objectives of the government and community concerns.

Essentially, the adoption of the 3D BIM-GIS models has increased the liability, sustainability, and inclusiveness of the urban spaces in Helsinki. Using digital twins and geospatial intelligence, the city is an excellent example of how data-based city regeneration can accommodate both the ecological, economic, and social levels, and serves as a foresightful precedent that can be applied to city planning around the globe. The Figure 5 shows the crucial results realized in Helsinki by the application of 3D BIM-GIS integration in urban regeneration projects are highlighted. It measures positive changes in five areas of planning priorities as follows: an 18% increase in horizontal space, 25% of redevelopment lands to be zoned green, a 22% decrease in commute time, a 30% increase in bicycle access and 60% citizen inclusion in planning. The chart emphasizes the role of the data-oriented and participatory approach adopted in Helsinki to encourage sustainable land use, seamless urban transport, and empower community, which proves the potential transformability of BIM-GIS synergy in the realization of inclusive and resilient smart cities.

Figure 5. Impact of BIM-GIS Integration in Helsinki's Urban Regeneration

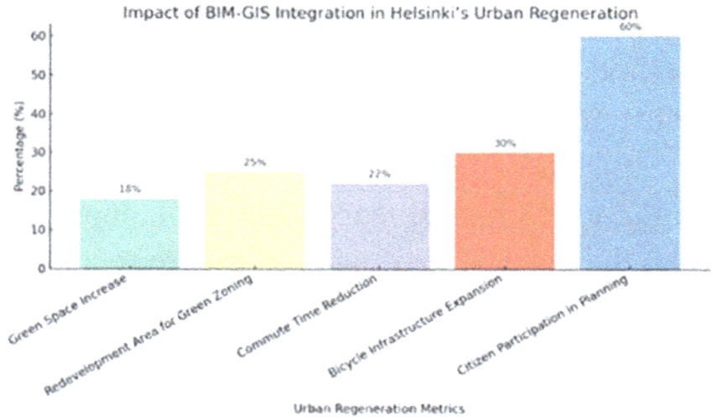

9. CONCLUSION

Building Information Modeling (BIM) coupled with Geographic Information Systems (GIS) is transforming the future of cities, infrastructure planning and operation of smart cities. In this Chapter, it was revealed how the BIM-GIS convergence is the heart of the real-time simulation of the city and implementing Digital Twin-based applications, as it makes it move beyond the frame of the traditional usage of siloed and static data management to building the dynamic, integrative, and intelligent urban environment. With this combination of micro-level information contained in BIM and macro-level application of spatial intelligence of GIS, urban planners, engineers and policy makers are in a position to model, analyze and manage the built environment and the whole natural environment, never with such precision. With the adoption of this integration, sustainable development is encouraged through performance simulation, green zoning and climate-adaptive planning. Spanning the gamut of maximizing the energy efficiency of buildings and decreasing their carbon footprints to green infrastructure mapping and smart energy networks, the BIM-GIS synergy allows the planners to plan based on environmental resilience. The integrated solution supports lifecycle infrastructure assets, predictive maintenance and real-time view of operations operation to guarantee urban utilities and service lifetime, safety and efficiency.

The practical uses of BIM-GIS integration found in the case studies of Singapore and Helsinki are the references to the development of smart infrastructure and urban regeneration. With 85% coverage on underground infrastructure mapping done digitally in the Marina Bay in Singapore, project inefficiencies such as the %age of project delays were reduced by 30% together with enhanced coordination between the parties involved in the project. The practice of 3D models by the city of Helsinki in participatory planning has led to the active involvement of 60% of the population in the process, and a 30% increase in the bicycle infrastructure network-the social and mobility benefits of data-driven urban change, in one of the cities of Europe. In addition, new technologies such as IoT or AI and edge computing are increasing the capabilities of BIM-GIS systems. The real-time sensor data coupled with computing at the periphery, analytics, and AI, will enable hands-free functioning, autonomous decision making, and real-time simulations. These technologies are bringing about the development of more sophisticated digital twins, which not only mirror the current state of a city but are also capable of predicting and optimizing future performances. Intelligent digital twins will also facilitate future scenario sifting and decision making. They will cut the time needed for scenario simulations by several orders of magnitude and allow evaluation of hundreds of thousands of options in real-time. These projections will be fault tolerant and come with a clear explainability of their decision paths.

Data availability silos, the absence of a common architecture, and political fragmentation will still enable the foundational pillars. Governance and digital infrastructure and the fairness of access down to the last mile will enable longitudinal sustainability and paralleling success in a competitive system. Policies should be intersectional and formed in strong collaborations between the builders of the smart eco systems on the ground and the digital policymakers. These systems should be designed with transparency, inclusive, and ethical consideration which also contributes to the enhancement of the urban system itself. In the busy pace of global urbanization, it is essential to adopt the synergistic properties of the BIM-GIS framework as the future cities are to be sustainable, resilient, and smart. The advent of the future rests unequivocally on the culminative integration of the above-denoted partnerships, aseptic protocols, and fused percipience unto the advancement of urbane spheres and resuscitative eco-friendliness.

10. Key Terms and Definitions

The Table 3 below introduces the key terminology that would be utilized throughout the chapter, so the readers had a clear idea of the fundamental concepts behind such technologies as Building Information Modeling (BIM), Geographic Information Systems (GIS), and Digital Twin technologies. It aids in understanding technical language, situational citations in the design, development and implementation of smart cities, infrastructure management and spatial intelligence.

Table 3. Key Terms and Definitions in BIM-GIS Integrated Urban Planning

Term	Definition
BIM (Building Information Modeling)	A digital representation of physical and functional characteristics of a building or infrastructure, providing detailed geometry, materials, and lifecycle data for construction and facility management.
GIS (Geographic Information Systems)	A spatial system used to capture, store, manipulate, analyze, manage, and present geographical data for land, terrain, and environmental analysis.
Digital Twin	A dynamic, virtual model of a real-world object, system, or city that is updated in real time using sensor data, enabling simulation, monitoring, and prediction.
3D City Modeling	The process of creating detailed, three-dimensional digital representations of urban environments, including buildings, roads, vegetation, and terrain.
Interoperability	The ability of different systems, data formats, and software platforms (e.g., BIM and GIS) to work together, exchange data, and interpret shared information effectively.
CityGML	An open data model and XML-based format for storing digital 3D models of cities and landscapes, used primarily in GIS.
IFC (Industry Foundation Classes)	A standardized, open file format used in BIM to describe building and construction industry data, ensuring interoperability across software.
Smart City	An urban area that uses digital technologies, data analytics, and intelligent systems to enhance operational efficiency, sustainability, and quality of life for residents.
Participatory Planning	An inclusive urban planning approach that involves community stakeholders in decision-making processes through tools such as 3D visualization, simulations, and feedback mechanisms.
Edge Computing	A computing paradigm that processes data closer to the data source (e.g., IoT devices) to reduce latency and bandwidth, essential for real-time urban applications.

11.REFERENCES

Andrianesi, D., & Dimopoulou, E. (2020). An integrated bim-gis platform for representing and visualizing 3d cadastral data. Isprs Annals of the Photogrammetry Remote Sensing and Spatial Information Sciences, VI-4/W1-2020, 3-11. DOI: 10.5194/isprs-annals-VI-4-W1-2020-3-2020

Avezbaev, S., Avezbaev, O., Tashpulatov, S., & Sharipov, S. (2023). Implementation of gis-based smart community information system and concepts of digital twin in the field of urban planning in uzbekistan. E3s Web of Conferences, 386, 05006. DOI: 10.1051/e3sconf/202338605006

Basir, W., Ujang, U., & Majid, Z. (2023). Adaptation 4d and 5d bim for bim/gis data integration in construction project management. *IOP Conference Series. Earth and Environmental Science, 1274*(1), 012002. DOI: 10.1088/1755-1315/1274/1/012002

Basir, W., Ujang, U., & Majid, Z. (2023). Adaptation 4d and 5d bim for bim/gis data integration in construction project management. *IOP Conference Series. Earth and Environmental Science, 1274*(1), 012002. DOI: 10.1088/1755-1315/1274/1/012002

Batty, M., Axhausen, K. W., Giannotti, F., Pozdnoukhov, A., Bazzani, A., Wachowicz, M., Ouzounis, G., & Portugali, Y. (2012). Smart cities of the future. *The European Physical Journal. Special Topics, 214*(1), 481–518. DOI: 10.1140/epjst/e2012-01703-3

Boccardo, P., Riccia, L., & Yadav, Y. (2024). Urban echoes: Exploring the dynamic realities of cities through digital twins. *Land (Basel), 13*(5), 635. DOI: 10.3390/land13050635

Borkowski, A., Kochański, Ł., & Wyszomirski, M. (2022). A case study on building information (bim) and land information (lim) models including geospatial data. *Geomatics and Environmental Engineering, 17*(1), 19–34. DOI: 10.7494/geom.2023.17.1.19

Cao, Y., Xu, C., Aziz, N., & Kamaruzzaman, S. (2023). Bim–gis integrated utilization in urban disaster management: The contributions, challenges, and future directions. *Remote Sensing (Basel), 15*(5), 1331. DOI: 10.3390/rs15051331

Colucci, E., Iacono, E., Matrone, F., & Ventura, G. (2023). The development of a 2d/3d bim-gis web platform for planned maintenance of built and cultural heritage: the main10ance project. The International Archives of the Photogrammetry Remote Sensing and Spatial Information Sciences, XLVIII-M-2-2023, 433-439. DOI: 10.5194/isprs-archives-XLVIII-M-2-2023-433-2023

Ding, X., Yang, J., Liu, L., Huang, W., & Wu, P. (2020). Integrating ifc and citygml model at schema level by using linguistic and text mining techniques. *IEEE Access : Practical Innovations, Open Solutions*, *8*, 56429–56440. DOI: 10.1109/ACCESS.2020.2982044

DURSUN. İ., Varlık, A., & Ayyıldız, E. (2022). Integration and web-based presentation of 3d city models and bim data: the case of the köyceğiz campus. The International Archives of the Photogrammetry Remote Sensing and Spatial Information Sciences, XLIII-B4-2022, 513-519. https://doi.org/DOI: 10.5194/isprs-archives-xliii-b4-2022-513-2022

Fang, J., Xie, X., & Xu, J. (2024). Development and key technologies for architectural design and technical solutions of cim platform: A review. *IOP Conference Series. Earth and Environmental Science*, *1337*(1), 012036. DOI: 10.1088/1755-1315/1337/1/012036

Haveman, S., Peng, X., Zhang, R., & Rahola, T. (2024). Towards social digital twins – an integrated sociotechnical approach for the urban energy transition., 479-496. DOI: 10.18690/um.fov.4.2024.28

Hijazi, I., Ehlers, M., Zlatanova, S., Becker, T., & Berlo, L. (2010). Initial investigations for modeling interior utilities within 3d geo context: transforming ifc-interior utility to citygml/utilitynetworkade., 95-113. DOI: 10.1007/978-3-642-12670-3_6

Hijazi, I., Krauth, T., Donaubauer, A., & Kolbe, T. (2020). 3dcitydb4bim: a system architecture for linking bim server and 3d citydb for bim-gis-integration. Isprs Annals of the Photogrammetry Remote Sensing and Spatial Information Sciences, V-4-2020, 195-202. DOI: 10.5194/isprs-annals-V-4-2020-195-2020

Hor, A., Jadidi, A., & Sohn, G. (2016). Bim-gis integrated geospatial information model using semantic web and rdf graphs. *ISPRS Annals of the Photogrammetry, Remote Sensing and Spatial Information Sciences*, *III-4*, 73–79. DOI: 10.5194/isprs-annals-III-4-73-2016

Kang, T. (2023). Scan to bim mapping process description for building representation in 3d gis. *Applied Sciences (Basel, Switzerland)*, *13*(17), 9986. DOI: 10.3390/app13179986

Kiavarz, H., Jadidi, M., Rajabifard, A., & Sohn, G. (2018). Bim-gis oriented intelligent knowledge discovery. *The International Archives of the Photogrammetry, Remote Sensing and Spatial Information Sciences*, *XLII-4*(W10), 79–82. DOI: 10.5194/isprs-archives-XLII-4-W10-79-2018

Leng, S. and Zhen-Zhong, H. (2020). A lightweight bim-gis integration method for rural building design and construction., 25-30. DOI: 10.3311/CCC2020-071

Liu, X., Wang, X., Wright, G., Cheng, J., Li, X., & Liu, R. (2017). A state-of-the-art review on the integration of building information modeling (bim) and geographic information system (gis). *ISPRS International Journal of Geo-Information, 6*(2), 53. DOI: 10.3390/ijgi6020053

Maltseva, I., Meng, T., Xiao, S., Cao, Y., & Zeng, L. (2024). An integrated review and analysis of urban building seismic disaster management based on bim-gis. E3s Web of Conferences, 474, 02013. DOI: 10.1051/e3sconf/202447402013

Mekawy, M., Issa, M., & Perjons, E. (2021). Bim-gis integration for traffic safety in cities. Isprs Annals of the Photogrammetry Remote Sensing and Spatial Information Sciences, VIII-4/W2-2021, 161-166. DOI: 10.5194/isprs-annals-VIII-4-W2-2021-161-2021

Meschini, S., Accardo, D., Avena, M., Seghezzi, E., Tagliabue, L., & Giuda, G. (2022). Data integration through a bim-gis web platform for the management of diffused university assets.. DOI: 10.35490/EC3.2022.217

Narindri, B., Nugroho, A., & Aminullah, A. (2022). Developing building management system framework using web-based-gis and bim integration. *Civil Engineering Dimension, 24*(2), 71–84. DOI: 10.9744/ced.24.2.71-84

Ohori, K., Diakité, A., Krijnen, T., Ledoux, H., & Stoter, J. (2018). Processing bim and gis models in practice: Experiences and recommendations from a geobim project in the netherlands. *ISPRS International Journal of Geo-Information, 7*(8), 311. DOI: 10.3390/ijgi7080311

Onungwa, I., Olugu-Uduma, N., & Shelden, D. (2021). Cloud bim technology as a means of collaboration and project integration in smart cities. *SAGE Open, 11*(3), 21582440211033250. Advance online publication. DOI: 10.1177/21582440211033250

Sani, M., Musliman, I., & Rahman, A. (2022). Ifc to citygml conversion algorithm based on geometry and semantic mapping. The International Archives of the Photogrammetry Remote Sensing and Spatial Information Sciences, XLVI-4/W3-2021, 287-293. DOI: 10.5194/isprs-archives-XLVI-4-W3-2021-287-2022

Sladić, G., Milosavljević, B., Nikolic, S., Sladić, D., & Radulović, A. (2021). A Blockchain Solution for Securing Real Property Transactions: A Case Study for Serbia. *ISPRS International Journal of Geo-Information, 10*(1), 35. DOI: 10.3390/ijgi10010035

Song, Y., Wang, X., Tan, Y., Wu, P., Sutrisna, M., Cheng, J., & Hampson, K. (2017). Trends and opportunities of bim-gis integration in the architecture, engineering and construction industry: A review from a spatio-temporal statistical perspective. *ISPRS International Journal of Geo-Information*, *6*(12), 397. DOI: 10.3390/ijgi6120397

Tan, Y., Liang, Y., & Zhu, J. (2023). Citygml in the integration of bim and the gis: Challenges and opportunities. *Buildings (Basel, Switzerland)*, *13*(7), 1758. DOI: 10.3390/buildings13071758

Tobiáš, P. (2016). Bim, gis and semantic models of cultural heritage buildings. *Geoinformatics Fce Ctu*, *15*(2), 27–42. DOI: 10.14311/gi.15.2.3

Usmani, A., Jadidi, M., & Sohn, G. (2021). Towards the automatic ontology generation and alignment of bim and gis data formats. Isprs Annals of the Photogrammetry Remote Sensing and Spatial Information Sciences, VIII-4/W2-2021, 183-188. DOI: 10.5194/isprs-annals-VIII-4-W2-2021-183-2021

Vacca, G., Quaquero, E., Pili, D., & Brandolini, M. (2018). Integrating bim and gis data to support the management of large building stocks. *The International Archives of the Photogrammetry, Remote Sensing and Spatial Information Sciences*, *XLII-4*, 647–653. DOI: 10.5194/isprs-archives-XLII-4-647-2018

Wang, B. (2021). The seductive smart city and the benevolent role of transparency. [s]. *Interaction Design and Architecture*, (48), 100–121. DOI: 10.55612/s-5002-048-005

Zhao, L., Liu, Z., & Mbachu, J. (2019). Highway alignment optimization: An integrated bim and gis approach. *ISPRS International Journal of Geo-Information*, *8*(4), 172. DOI: 10.3390/ijgi8040172

Zhao, L., Liu, Z., & Mbachu, J. (2019). An integrated bim–gis method for planning of water distribution system. *ISPRS International Journal of Geo-Information*, *8*(8), 331. DOI: 10.3390/ijgi8080331

Zhu, J., Wright, G., Wang, J., & Wang, X. (2018). A critical review of the integration of geographic information system and building information modelling at the data level. *ISPRS International Journal of Geo-Information*, *7*(2), 66. DOI: 10.3390/ijgi7020066

Zhu, J., & Wu, P. (2021). Towards effective bim/gis data integration for smart city by integrating computer graphics technique. *Remote Sensing (Basel)*, *13*(10), 1889. DOI: 10.3390/rs13101889

Zhu, J., Wu, P., & Anumba, C. (2021). A semantics-based approach for simplifying ifc building models to facilitate the use of bim models in gis. *Remote Sensing (Basel)*, *13*(22), 4727. DOI: 10.3390/rs13224727

Zhu, J., Wu, P., Chen, M., Kim, M., Wang, X., & Fang, T. (2020). Automatically processing ifc clipping representation for bim and gis integration at the process level. *Applied Sciences (Basel, Switzerland)*, *10*(6), 2009. DOI: 10.3390/app10062009

Chapter 3
Digital Twins for Green Infrastructure:
Extending BIM and GIS Integration

Ulugbek Eshqarayev
Termez University of Economics and Service, Uzbekistan

Shokhjakhon Akhmedov
Urgench State University, Uzbekistan

Baxtiyor Egamov
Mamun University, Uzbekistan

Munir Ahmad
https://orcid.org/0000-0003-4836-6151
Survey of Pakistan, Pakistan

ABSTRACT

This chapter examines the transformative potential of integrating Digital Twins with BIM and GIS for green infrastructure planning and management. By combining real-time data, spatial intelligence, and performance modeling, Digital Twins enable dynamic, system-wide insights across diverse typologies such as green roofs, wetlands, and urban parks. While offering powerful tools for climate-resilient and sustainable urban development, adoption remains limited by technical, institutional, and regulatory challenges. The chapter outlines strategic directions for research, policy, and capacity-building to scale ethical, interoperable, and inclusive Digital Twin solutions for future-ready, green cities.

DOI: 10.4018/979-8-3373-4058-6.ch003

INTRODUCTION

The term green infrastructure encompasses a network of strategically planned natural and semi-natural systems that provide a wide range of benefits to the environmental, social, and economic conditions, including parks, wetlands, green roofs, urban forests, and water-sensitive features, among others. Green infrastructure is an important element of urban resilience, environmental quality, and community well-being as urban centers experience increasing pressure from climate change, resource loss, and population increase. The difference between green solutions and traditional grey infrastructure lies in the fact that green design incorporates ecological processes into cities so that these cities can ameliorate floods, avoid heat islands, clean air and water, and offer recreational areas. Nonetheless, it needs to deal with these complex dynamic systems with innovative planning, monitoring, and optimization efforts (Fang et al., 2023; Green et al., 2021; Lopes et al., 2025) .

The main aim of achieving all the potential of the green infrastructure would require the cities to fuse a long-time fragmentation of how the data would be collected, analyzed, and realized within planning and operational systems. BIM entered the modern picture with the ability to create detailed descriptive information about physical structures and material characteristics that are object-oriented. Geographic Information Systems (GIS) offer the facility of modeling the environment and conducting spatial analytical work in the wider scope of urban design. A combination of BIM and GIS can provide a more comprehensive picture of both the city infrastructure above ground and underground infrastructure, but the development of Digital Twins goes a step further by integrating real-time data feeds, predictive models, and interactivity with simulation into live virtual models. Digital Twins provide a real-time reflection of the real-world assets, where they mimic the environmental conditions and generate performance optimization and proactive actions. They are particularly useful when it comes to green infrastructure, in which both natural variability and human use patterns are to be heeded in lifecycle management (Babalola et al., 2023; T. Tan et al., 2021; Xia et al., 2022).

This chapter discusses the transformative power of Digital Twins in the design, operationalization, and optimization of green infrastructure systems based on the use of a combined framework of BIM and GIS. It seeks to give a knowledgeable view of the technical architecture supporting Digital Twins, which involves data interoperability standards, sensor integration, and the ability to simulate. In this chapter, we shall demonstrate how Digital Twins would be utilized to optimize the urban green spaces, water-sensitive infrastructure, and climate-resilient systems. Moreover, it will highlight major implementation challenges (including data governance, institutional coordination, and others) and suggest facilitating strategies in the effort to accommodate scalable adoption. The concentration on the intersection of digital technologies with environmental planning and planning can facilitate the placement of Digital Twins and Digital Twins in the context of sustainable, data-driven urban development at the forefront.

BIM-GIS CONVERGENCE

Geographic Information Systems (GIS) and Building Information Modeling (BIM) are two of the foundational technologies in the process of planning, designing, and managing both built environments and natural systems. BIM concentrates on micro-level modeling of physical structures and provides detailed, object-oriented 3D models with rich data that embraces a broad scope of architectural, structural, mechanical, and operational information. The geometry, properties, performance parameters, and interdependencies of each component, be it a wall, HVAC unit, or foundation slab, are digitally encoded, and thus precise simulations, clash detection, construction sequencing, and facility management throughout the asset lifecycle are possible. BIM is prevalently used in areas of vertical construction, infrastructure engineering, and asset management because of its accuracy and relative detail (Ali et al., 2022; Durdyev et al., 2022; Olanrewaju et al., 2022; Vignali et al., 2021).

Conversely, GIS offers macro-based spatial intelligence that combines data collection instruments and analysis, and mapping applications to be able to better plan land-use, environmental analysis, utility mapping, demographic analysis, and infrastructure network optimization. GIS is very useful in terrain models, hydrology data (Thakur et al., 2017), zoning, transport network (Önden et al., 2024), vegetation index (Islam et al., 2023; Senhorelo et al., 2024), and climate-based data (Alahacoon & Edirisinghe, 2021; Todeschi et al., 2022), and thus GIS is crucial in regional planning, ecological modeling, disaster forecasting, and assessment. It is strong in terms of using georeferencing, spatial overlay, and network analytics with large geographic scopes, such as neighborhoods to large geographic areas like metropolitan regions.

BIM and GIS have long existed in a parallel way, despite their original features, regardless of their ability to cross-communicate with one another. The BIM system works mainly under local coordinate systems, valuing engineering precision and constructive common sense, and the GIS system works in the global coordinate system with a concern for the environment and spatial interrelationship. This technical and disciplinary separation has limited the capacity of cities and infrastructure managers to come up with integrated, intelligent, and context-sensitive systems that can take into consideration built form and environmental functionality concomitantly. BIM and GIS are, as such, not only converging in a technical necessity but benefiting in the strategic capacity to see the fruition of smarter, more resilient, and sustainable infrastructure, especially in areas such as green infrastructure, where natural processes and engineered systems are closely interconnected.

Digital Twins

Digital Twins are a paradigm of the same reason that we view, implement, and transform infrastructure systems, especially infrastructure systems that are constituted in dynamic environmental settings. Digital Twins are the logical extension of BIM and GIS integration as they equip the static model with dynamic data-driven and actively synchronized virtual representation of physical assets, systems, or the overall environment (Avezbaev et al., 2023; Lee & Lee, 2021; Tripathi et al., 2023; Xia et al., 2022; Zhu & Wu, 2021). Such virtual representations are energized by streams of real-time data providing information generated by sensors, Internet of Things devices, weather stations, satellite feeds, and operations management systems. This interconnectedness allows a two-way interaction wherein any changes in the real world are instantly updated in the virtual one-and the same can be said in the reverse, that the digital computer models can be used to guide physical actions.

Contrary to conventional BIM and GIS models, which have indeed their purpose limited to design, planning, and archiving, the Digital Twin solutions are used in subsequent phases, operational and adaptive, of the asset lifecycle. They enable a large and diverse range of high-value uses: predictive maintenance profiting off machine learning to anticipate failure in advance of it happening; scenario testing potential site setups in simulations; environmental modeling, even simulating microclimates or hydrological flows; and behavioral analytics to optimize user experience in shared green spaces. Digital Twins, together with the geometric model precision and material-level intelligence of BIM, with the spatial context and environmental modeling of GIS, enable the combined framework to represent both the internal behaviour of the infrastructure and its relationship to neighbouring ecosystems.

It is especially transformative in the context of green infrastructure (Bianchini et al., 2023; Ozel & Petrovic, 2024). They may simulate the impact on building temperature over time of a green wall, calculate the stormwater absorption in rain gardens in times of extreme weather, or estimate the long-term effect of urban forestry on air quality and biodiversity. Further, planners, engineers, and policymakers can use these digital settings to co-create the design and real-time modification of green systems with references to the histories of these systems as well as projections of the systems. The fact that Digital Twins can be viewed as living systems, in the sense of being capable of learning, adapting, and optimizing performance, makes them some of the critical tools in creating resilient, climate-responsive, and sustainable urban infrastructure. Figure 1 illustrates the conceptual evolution from traditional GIS and BIM systems toward the more dynamic and integrated framework of Digital Twins, emphasizing their increasing complexity, data integration, and real-time capabilities.

Figure 1. Conceptual Evolution from GIS and BIM to Digital Twins

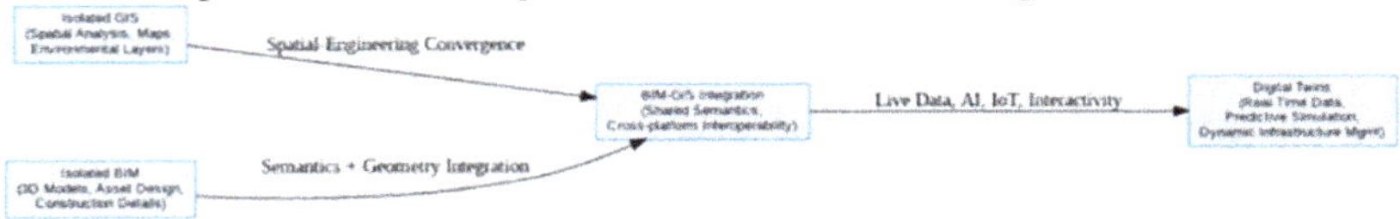

TECHNICAL ARCHITECTURE OF DIGITAL TWINS FOR GREEN INFRASTRUCTURE

Layered Architecture

Digital Twin of green infrastructure can be structured into four data-dependent tiers: (1) the data suppliers, (2) modeling and analytics engines, (3) integration and interoperability structures, and (4) visualization and interaction structures. Such a hierarchical structure enables the free exchange of data between the physical systems and the digital platforms and vice versa to perform a continuous tracking, simulation, and optimization of the entire process.

The most fundamental one is the data source layer that gathers various sources of information. These are geospatial databases (e.g., land cover, soil type, hydrology), BIM models (e.g., drainage detail and structural systems), Internet of Things (IoT) sensors (e.g., soil moisture, air quality, and flow rates), satellite imagery, drone data, and citizen-reported observations. Such data flows are consumed by the modeling and analytics layer that acts as the powerhouse of the Digital Twin. On this tier, there are reruns of simulations, AI models, hydrological or energy simulations, and dashboards to monitor performance and support real-time decision-making and predictive analytics.

The integration frameworks layer does this, the third layer, bringing together inputs in terms of domains into a harmonic/balanced position to achieve semantics and spatial / time consistency. It is a linkage between GIS and BIM environments that allows operations to flow both ways and convert objects across operating systems. The visualization layer delivers synthesized information to the users in an easy-to-understand 2D map, 3D models, dashboards, or augmented/virtual reality interfaces. This layer facilitates an immersive experience among the engineering, urbanization, and community citizens to allow transparency and their participation in design.

Data Interoperability Standards

The major enabler of such a layered architecture is the commitment to interoperability standards that allow smooth interchange of information and its interpretation across systems. The principal standards are:

Industry Foundation Classes (IFC)

IFC enables a consistent mechanism for modeling the geometry, spatial, material, mechanical system, and lifecycle characteristics of building and infrastructure projects (Liu et al., 2021; Okonta et al., 2024). IFC facilitates teamwork between architects, engineers, contractors, and facility managers to work together more efficiently, and also allows them to move beyond the proprietary descriptions. Under its Digital Twins of green infrastructure, IFC may act as the foundation on top of which highly detailed information about building structures and systems, like the specifications of a green roof or stormwater retention tank, is imported into a more general digital environment that also looks beyond building-specific data (environmental and spatial data).

CityGML

CityGML is an open data model and XML format applied in GIS to the 3D format and semantics of the urban space (Kutzner et al., 2020; Y. Tan et al., 2023). Conceived by the Open Geospatial Consortium (OGC) and used in smart cities, CityGML models different objects of the city, like the terrain, buildings, vegetation, water bodies, and transportation infrastructure, in diverse levels of detail based on simple block models to detailed construction types. The difference with CityGML is that it is semantically rich; it not only represents geometry, but also stores semantics in the form of thematic attributes, such as building use, roof type, or vegetation type. CityGML facilitates spatial modeling of both natural constructions (e.g., green corridors, bioswales) and built parts of green infrastructure, and thus makes it possible to model ecological performance of green infrastructure and urban-climate dynamics scenarios within Digital Twins.

InfraGML

InfraGML is a profile of the Geography Markup Language (GML), based on civil engineers and infrastructure data interchange requirements (Kumar et al., 2019a; Open Geospatial Consortium, 2017). It can give systematic representations of transportation networks, drainage systems, utilities, and other linear facilities of infrastructure, and is suitable for large projects like highway, bridge, stormwater, projects. InfraGML presents alignment geometry, surface models, and asset metadata, and is geospatially aware, and is therefore interactable with CAD and GIS tools. InfraGML allows detailed and standardized modeling of such systems as permeable pavement layouts or culvert connections in a Digital Twin of green infrastructure, which in turn lends itself to both engineering design precision and environmental analysis.

LandInfra

LandInfra is a shared international standard developed by the Open Geospatial Consortium (OGC) and ISO together to facilitate the representation and integration of land administration and civil engineering data as relevant to geospatial environments (Buuveibaatar et al., 2022; Kumar et al., 2019b). It offers a comprehensive plan for overseeing survey data, retainers of acre and easements alongside civil infrastructure objects in a congruent geospatial situation. LandInfra is filling in that gap between traditional engineering data models and GIS because it supports things like land ownership boundaries, rights-of-way, and even asset alignments. LandInfra, in the context of green infrastructure, enables integration of legal, cadastral, and physical infrastructure data, which is especially useful in planning of nature-based solutions in highly urban land administration systems. The standards guarantee that the spatial and structural models can reside in the same digital setting and contribute to software interoperability at the level of such software products as Revit, ArcGIS, QGIS, etc., or open-source solutions like FME or BIMServer. This is essential to green infrastructure, e.g. combination of BIM-based storm water detention tank designs and CityGML-based hydrological terrain model provides a common view of flood risk and capacity design.

Real-time Data Pipelines and Sensor Integration

Integration with the sensors is the blood of an operational functioning Digital Twin, which turns a static model into a dynamic, reactive, real-time model. When applied to green infrastructure, an extensive number of all types of environmental sensors can be commissioned.

Soil Moisture and Temperature Probes

Important operating elements to observe the hydrological and thermal performance of bioswales, rain gardens, and other vegetated stormwater management systems are soil moisture and temperature probes (Kashyap & Kumar, 2021; Rasheed et al.,

2022). With these sensors, locations across the landscape can be monitored with volumetric water content and soil temperature data in different depths, and provide real-time measurements of infiltration rates, evapotranspiration processes, and root zone conditions in plants. In Digital Twins of green infrastructure, this information can be used to calibrate the hydrological models, get an idea of the system performance in conditions of precipitation, and use adaptive irrigation schemes to ensure maximum vegetation growth and water absorption capacity.

Weather Stations

Precipitation levels, solar irradiance, temperature, humidity, wind speed, and direction are just a few of the more critical environmental inputs that are obtained through the use of on-site or networked weather stations (Kampczyk & Dybeł, 2021). These parameters are specifically applicable in the simulation of the generation of runoff, evapotranspiration, and microclimate impacts of green infrastructure, such as green roofs, rain gardens, and urban forests. Within a Digital Twin, weather station data facilitates dynamic environmental modeling, facilitates optimization of the performance of vegetation, predicts the system response to varying climatic conditions, and to what extent under extreme weather events.

Air Quality Sensors

A growing number of air quality sensors are used in and around green corridors and urban forests to assess their environmental co-benefits, measure pollutants such as PM2.5, PM10, and ozone (Popoola et al., 2018; Zhang & Srinivasan, 2020). These sensors record the effect that vegetation has on the quality of air around by both filtering out particulates and by trapping gaseous pollutants. This sensor data could be utilized to simulate spatial air quality improvements when combined with a Digital Twin, could calculate human health benefits, and be used to optimize where or the species mix of urban green spaces can be placed with respect to pollution reduction requirements.

Flow Meters and Level Sensors

The performance of permeable pavements, detention basins, and Stormwater retention tanks needs flow meters and level sensors to ensure their monitoring (Wang et al., 2021). These sensors indicate the rates of water flow, the amount of water stored, and the volume of discharged water in real time, therefore giving a proper picture of the efficiency of stormwater capturing, the rate of infiltration, and the possibility of an overflow. Within a Digital Twin, such data is input into predictive algorithms to model the behavior of the drainage system at different levels of rainfall and facilitate a proactive approach to management of a stormwater network under peak conditions or flow events.

Vegetation Health Sensors

Drone or satellite imagery produces the Normalized Difference Vegetation Index (NDVI), which is (Madeira et al., 2022)commonly known to evaluate the vegetation health and biomass under any conditions of the urban green infrastructure properties. NDVI is a measure of chlorophyll activity, so it gives signs of plant health, stress levels, and spatial variation in the condition of the ecosystem. The NDVI information can be used in Digital Twins to dynamically display how certain areas are performing, remote diagnostic of stressed areas, and inform management decisions of whether certain areas need maintaining or replanting, or estimation of ecosystem services; e.g., the capacity to sequester carbon or cool cities (Huang et al., 2021; Tunio et al., 2025).

The data traveling through IoT gateways and edge devices is sent by these sensors to cloud-based platforms or the local control systems, which is done with the preprocessing of the data in the real-time data pipelines (via MQTT or OPC-UA protocol), among others. Ingestion, routing, and security are handled by middleware platforms, such as Azure Digital Twins, AWS IoT Core, or open-source alternatives like FIWARE. Such pipelines can be directly connected to analytics engines and visualization dashboards to enable dynamic reaction, e.g., turning irrigation systems on or off, notifying users of the risk of overrunning a system, or acting on infrastructure behavior (e.g., changing its use patterns under climate feedback). Figure 2 outlines the technical architecture of a Digital Twin tailored for green infrastructure.

Figure 2. Digital Twin Architecture for Green Infrastructure

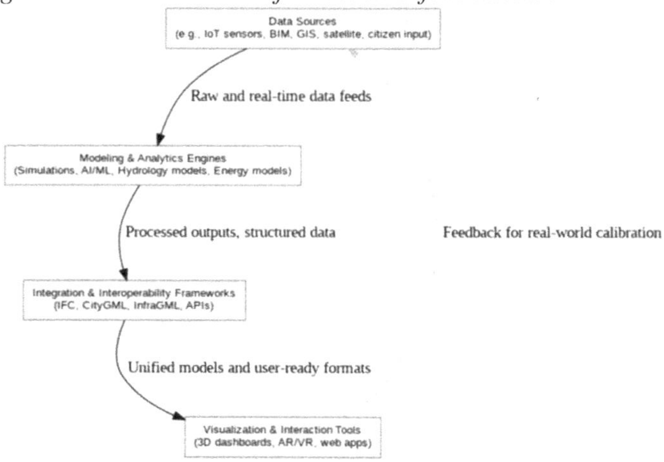

APPLICATIONS IN SUSTAINABLE URBAN SYSTEMS

Digital Twins are transforming the ideas of sustainability in constructing sustainable urban systems, by way of conceptualization, implementation, and maintenance. They combine real-time monitoring, predictive, and spatial intelligence into one dynamic platform that can handle the complexity of the green infrastructure in the ever-changing urban spaces. The section presents three domains of critical application areas under which Digital Twins provide a transformative value: Green space in the urban context, water-sensitive infrastructure, and a climate-resilient system. All applications are justified by real-life or conceptual case scenarios illustrating how BIM, GIS, and even real-time data can be integrated as ways to maximize environmental and social results.

Urban Green Spaces

City parks, green corridors, urban forests, green roofs, and other green areas are the main factors in bringing nature into cities in order to improve biodiversity, decrease stress levels in cities, and improve overall quality living. However, their maintenance gives problems linked to nature health, utilization, and sustainability. Digital Twins allow real-time city monitoring and simulation of urban green space performance by amalgamation of LiDAR-based canopy evaluation, NDVI vegetation health index, and drones or satellites, and public access data using sensor or mobile app (Gholami et al., 2022; Ozel & Petrovic, 2024).

To give an example, tracking the loss or gain of canopy cover over time, analysis of footfall patterns to identify underused areas, and a set of biodiversity indicators (bird song or pollinator activity) using acoustic sensors and image recognition algorithms can be monitored by a Digital Twin of an urban park. Singapore city has conceptually experimented with a Digital Twin framework of green corridors that integrates a geographical information system (GIS)-based spatial planning with real-time sensors of air quality and canopy to direct maneuvers of planting decisions and constructing biodiversity.

Water-Sensitive Infrastructure

To enhance percolation, decrease overflow, and filter disease-carrying agents, bioswales, rain gardens, retention basins, and permeable pavement have become more common in designs employed as part of stormwater management. Management skills of such systems demand the knowledge of the hydrological performance of the systems under different weather conditions, as well as under diverse urban loading conditions.

Digital Twins enable the use of smart drainage systems by consolidating level sensors, flow meters, weather station information, and hydrological models to make predictive modeling (Fan et al., 2025; Zhao et al., 2025). As one illustration, sensors integrated into permeable pavement systems can provide run-on infiltration and saturation information to a centralized twin in real-time and, in turn, dynamically adjust run-off potential and suggest maintenance activities such as debris removal or infiltration testing. A conceptual application in Melbourne, Australia, envisions a city-wide stormwater Digital Twin to model the flow within both the green and grey infrastructural backbones of the urban environment, facilitating predictive measures of flooding avoidance and future sustainable use of water via wetlands and green buffers.

Climate-Resilient Systems

The new infrastructure can need to foresee and respond to the growing number of extreme events, such as flash floods and urban heat waves. Digital Twins can be used proactively to simulate, test, and refine responses of cities to these challenges by combining climate projections, topographic models, and building information and sensor data into consistent real-time decision models.

With respect to flood resilience, Digital Twins have the potential to simulate surface runoff in flood areas, water buildup, and overflow across drainage basins by integrating rainfall sensor records, topographic, and drainage network models. The project of Digital Twin Flood Modeling project in Rotterdam allows simulating the behavior of water in the low-lying parts of the city on the basis of integration using sensor and spatial data, which allows preventing the risk in real time and organizing mutual assistance of first responders. Digital Twins have the potential to determine the temperature gradient around a city and evaluate how planting trees, providing reflective surfaces, or creating green roofs can reduce the temperature of particular areas as a solution to urban heat islands. By integrating thermal imagery with models of urban forms and climate data, it is possible to test out and optimize scenarios the most vulnerable people are at risk in cities. Table 1 provides a comparative overview illustrating how different green systems leverage Digital Twin technologies for monitoring, analysis, and decision-making.

Table 1. Use Case Matrix — Digital Twins Across Green Infrastructure Typologies

Infrastructure Type	**Key Sensors/ Inputs**	**Core Technologies**	**Digital Twin Functionality**	**Expected Outcomes**
Urban Park	NDVI, footfall, canopy LiDAR	GIS, IoT, AI Analytics	Biodiversity tracking, public use analysis	Optimized planting, equitable access
Permeable Pavement	Flow meters, soil probes	BIM-GIS Integration, Hydrological Models	Infiltration simulation, clog detection	Improved water retention, timely maintenance
Urban Wetland	Water level sensors, climate data	Remote Sensing, AI	Nutrient modeling, overflow prediction	Pollution control, habitat preservation
Green Roof	Temperature and humidity sensors	BIM, Energy Simulations	Heat stress analysis, structural moisture monitoring	Reduced UHI, building energy savings

CHALLENGES AND ENABLERS FOR IMPLEMENTATION

Although the prospect of Digital Twins in green infrastructure is immense, their deployment is limited by numerous barriers of technical, organizational, and policy-level barriers. It takes a mix of empowering technologies, cross-functional planning, and compliance with the principles of open standards and data governance to overcome these obstacles. The section investigates the major barriers and facilitators influencing the implementation environment of Digital Twins in sustainable urban infrastructure.

Technical Challenges

The most outstanding technical deterrent is the existence of data silos within fields. In spite of the fact that both BIM and GIS have their particular strengths, when applied exclusively, they both pose significant limitations that are particularly relevant in terms of green infrastructure planning, execution, and continual maintenance. In spite of being very useful in modeling the inner characteristics of buildings and of the components in the infrastructure, BIM is not environmentally situationally aware. It is defined within a project-based logic often limited by the footprint of an individual building or asset and normally not including more global ecological and geospatial contexts, likes as water connectivity, temperature fluctuation within cities, or wildlife networks. Consequently, some decisions made in the context of BIM environments can disregard important interactions between the environmental and urban systems.

GIS, on the other hand, gives a complete spatial analysis and environmental modelling on a greater scale than one level, portraying visibility of the terrain, drainage, land use, and vegetation layout; however, it does not include the detailed structural or mechanical information to describe or operate individual-built assets within a particular built asset. An example of this is that a GIS might be able to tell the flood-prone areas (Mourato et al., 2023; Mujib et al., 2021)or be able to map green corridors (Gargiulo et al., 2020; Rodríguez-Espinosa et al., 2020; Yan & Tang, 2021), but it does not know how well a building with its green roof should perform on a moderate rainy day, or how the underground storm water tanks can perform with the surrounding infrastructure.

That misrepresentation between the built form and digital depiction of the environmental context cripple's transcendental cooperation between the members of the architectural support team, including engineers, planners, ecologists, and facility managers. It also causes discontinued work patterns, redundant work, and less-than-ideal design solutions. The combination of both BIM and GIS serves this gap as the two systems allow a bi-directional flow of geospatial, semantic, and time information. It enables a more encompassing interpretation of the infrastructure operation beyond both a self-contained system and a component of an upper network of urban ecology. Such integration is particularly crucial concerning green infrastructure, including bioswales, green facades, permeable pavements, and urban wetlands. It allows an ecological performance to be simulated, the cumulative environmental impact to be measured, and to dynamically adapt to climatic variables. BIM-GIS integration with spatial intelligence can provide stakeholders with the power to produce infrastructure that is not only technically sound but also ecologically sensitive. This synthesis is a necessary basis for the achievement of resilient cities in which humans and the physical environment co-develop (Avezbaev et al., 2023; Breunig et al., 2020; Meschini et al., 2022).

Organizational Challenges

In an organizational sense, numerous local governments, planning agencies, and utilities do not have the necessary technical and cross-disciplinary coordination skills to create and maintain Digital Twin ecosystems. Employees can be well-versed in GIS or other tools engineering and not know how to combine BIM data, streaming IoT pipes, or simulation AI blending. Moreover, the division of institutional mandate with parks handled by one department, a water system by another, roads by a third, and ICT infrastructure by another gives bureaucratic friction to coming up with shared, cross-sectoral platforms. It is also problematic that there is no standard framework or business model for long-term maintenance, data sharing, and co-ownership of Digital Twins currently. As an example, who updates the twin in case of a new green asset being put in place? Lack of well-articulated governance frameworks makes most pilots fail to scale and endure past the implementation process.

Policy and Regulatory Challenges

In policy, there are no requirements for interoperability standards and regulatory incentives to promote the adoption of Digital Twin technologies in infrastructure procurement and urban planning. A wide range of current policies has no necessity to deliver in open formats (e.g., IFC, CityGML), which causes problems of vendor lock-in and inaccessible design data. There is also the problem of privacy and data safety since data on citizen movement or real-time environmental data may need to be merged; once this is done, there must be strict governance systems and anonymization processes. Procurement systems would also tend to capital cost rather than lifecycle performance, and simple and cheaper investment options fall upon dynamic digital tools capable of providing long-term value via optimization and resilience. To counter this, there must be a change in the policies of the public sector to practice performance-based infrastructure contracting and incorporate urban and environmental standards with requirements that are digital twin-ready.

Enabling Technologies

These obstacles notwithstanding, a number of emerging technologies can act as the enablers of the scalable deployment of the Digital Twin. The backbone of real-time feedback loops is IoT devices and sensor networks that enable cities to track the environmental conditions, the behaviour of the users, and the way the system works.

Digital Twins can be used to optimize infrastructure operations, detect anomalies, perform preventive maintenance, and other activities through artificial intelligence (AI) and machine learning models, thereby improving their analysis capabilities. As an example, AI can be used to study the prior data on rainfall and runoff to estimate the risks of overflow in green stormwater infrastructure. Also, Azure Digital Twins, AWS IoT Core, and Google Cloud IoT provide large-scale platforms to ingest, store, and process scales of time-series and spatial data. They frequently incorporate features such as digital twinset modeling, API-based integration, and visualization dashboards, which is an advantage to both the municipalities and the engineering firms in the context of barriers to entry.

Interoperability Frameworks and Open Data Principles

To achieve the potential of Digital Twins in green infrastructure fully, cities and organizations need to accept the promise of open data and interoperability standards. Appropriate standards like IFC, CityGML, LandInfra, and InfraGML enable the smooth transfer of BIM and GIS space, and SensorThings API and MQTT protocols allow standardization of real-time data transfer. Utilization of such structures helps to keep digital resources independent of a specific vendor, read by a computer, and not exposed to platform shifts in the future.

Moreover, open data policies, in which non-sensitive environmental and infrastructure data are published publicly, can speed innovation through the building of complementary services by third-party developers, researchers, and civic tech companies. Such a collaborative ecosystem method can also reduce the costs spent on the implementation of such green infrastructure, in addition to promoting transparency and co-creation in the planning and monitoring of such green infrastructure. Table 2 provides a list of challenges and enablers to implement digital twins for green infrastructure, whereas Figure 3 outlines a strategic decision-making process for adopting Digital Twins in green infrastructure projects, starting from data readiness to policy alignment.

Table 2. Challenges vs. Enablers in Implementing Digital Twins for Green Infrastructure

Challenge Domain	Key Challenges	Enabling Solutions
Technical	• Data silos across BIM, GIS, and IoT platforms • Model complexity and scalability • Lack of real-time integration	• Interoperability standards (IFC, CityGML, InfraGML) • Cloud-native platforms • SensorThings API, MQTT, OPC-UA
Organizational	• Skills gaps in cross-disciplinary tools • Fragmented responsibilities among departments • Lack of long-term digital asset governance	• Capacity-building programs • Digital Twin governance frameworks • Cross-sector working groups
Policy and Regulatory	• Absence of open data mandates • Vendor lock-in from proprietary formats • Weak digital procurement standards	• Open data policies • Performance-based procurement • Regulatory alignment with OGC/ISO standards
Technological Enablers	—	• IoT sensors for real-time feedback • AI/ML for predictive modeling • Cloud infrastructure (Azure, AWS, GCP)

Figure 3. Decision Framework for Digital Twin Adoption in Green Infrastructure

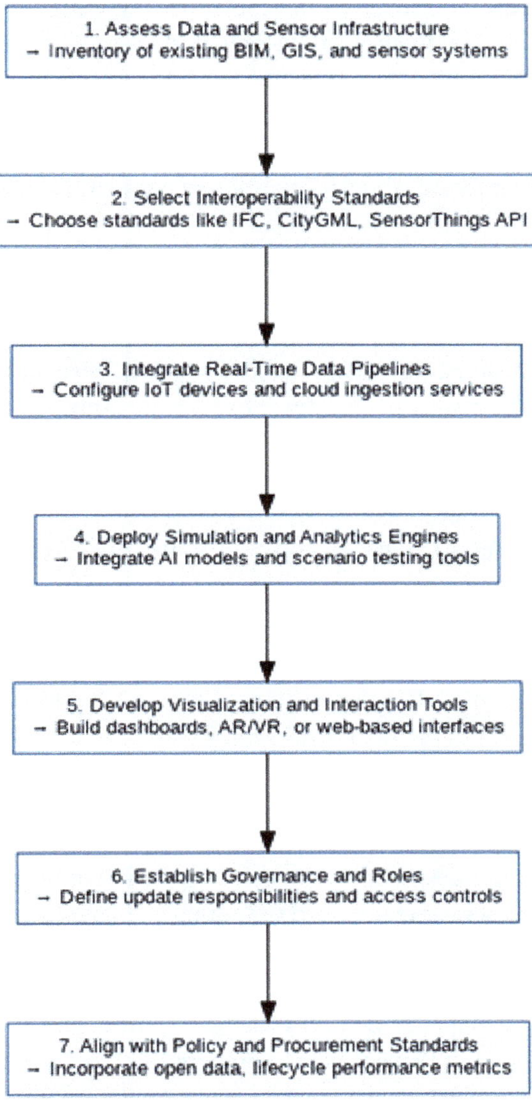

CONCLUSION

The combination of BIM and GIS with Digital Twins constitutes a breakthrough in terms of planning, implementing, and controlling green infrastructure. Digital

Twins allow a novel level of real-time monitoring and modeling at a scenario level, modeling the lifecycle optimization of said urban systems, due to the ability to present it in a form that is dynamic, data-rich, and spatially contextualized. In this chapter, how Digital Twins can improve the decision-making process across various typologies existing between urban parks and permeable pavements, wetlands, and green roofs was outlined due to their ability to interconnect environmental sensors, geospatial intelligence services, and infrastructure performance models in a coherent digital system.

Nevertheless, these innovations have come with significant constraints whose limitations are witnessed even when it comes to the implementation of Digital Twins in sustainable infrastructure that is mainly linked to issues such as technical fragmentation, organizational silos, skills gaps, as well as a lack of regulations that facilitate the exchange of open data and long-term digital governance. The elimination of these barriers can involve the cities investing in not just enabling technologies like IoT, AI, and cloud computing but also institutional capacities, workforce training, and cross-sectoral collaboration.

Moving forward, future study needs to look into the creation of uniform documentation on urban-natural systems, enhanced multi-scale-scenario foundation of simulation (e.g., microclimate and ecological relationships), and expanding lightweight, open-source Digital Twin frameworks applicable to smaller cities and transitional areas. It is also necessary to investigate further ethical aspects, including sovereignty of the data and environmental justice in the deployment of smart infrastructures.

At the policy level, governments need to emphasize the use of performance-based standards that specify the necessity to use digital twins' compatibility in new infrastructure projects and encourage the use of open and interoperable frameworks (e.g., IFC, CityGML, SensorThings API). Nothing can guarantee that Digital Twins develop as an instrument not of engineering, but of climate resilience, ecological recovery, and equal cities, then strategic alignment with smart city policies and green urban development plans. Digital Twins can be a persuasive means to envision new ways of infrastructure that are yet to come, not as fixed physical buildings and spaces, but as dynamic, responsive systems embedded within breathing urban life. By being leveraged in the right way as one of the foundation technologies, they can become a crucial element in the development of environmentally friendly, robust, and smart cities of the future.

REFERENCES

Alahacoon, N., & Edirisinghe, M. (2021). Spatial variability of rainfall trends in sri lanka from 1989 to 2019 as an indication of climate change. *ISPRS International Journal of Geo-Information, 10*(2), 84. Advance online publication. DOI: 10.3390/ijgi10020084

Ali, K. N., Alhajlah, H. H., & Kassem, M. A. (2022). Collaboration and Risk in Building Information Modelling (BIM): A Systematic Literature Review. *Buildings (Basel, Switzerland), 12*(5), 571. Advance online publication. DOI: 10.3390/buildings12050571

Avezbaev, S., Avezbaev, O., Tashpulatov, S., & Sharipov, S. (2023). Implementation of GIS-based Smart Community Information System and concepts of Digital Twin in the field of urban planning in Uzbekistan. *E3S Web of Conferences, 386*. DOI: 10.1051/e3sconf/202338605006

Babalola, A., Musa, S., Akinlolu, M. T., & Haupt, T. C. (2023). A bibliometric review of advances in building information modeling (BIM) research. *Journal of Engineering. Defence Technology, 21*(3), 690–710. DOI: 10.1108/JEDT-01-2021-0013

Bianchini, S., Damioli, G., & Ghisetti, C. (2023). The environmental effects of the "twin" green and digital transition in European regions. *Environmental and Resource Economics, 84*(4), 877–918. Advance online publication. DOI: 10.1007/s10640-022-00741-7 PMID: 36532702

Breunig, M., Bradley, P. E., Jahn, M., Kuper, P., Mazroob, N., Rösch, N., Al-Doori, M., Stefanakis, E., & Jadidi, M. (2020). Geospatial data management research: Progress and future directions. *ISPRS International Journal of Geo-Information, 9*(2), 95. Advance online publication. DOI: 10.3390/ijgi9020095

Buuveibaatar, M., Lee, K., & Lee, W. (2022). Development of a Conceptual Data Model for 3D Geospatial Road Management Based on LandInfra Standard: A Case Study of Korea. *ISPRS International Journal of Geo-Information, 11*(5), 316. Advance online publication. DOI: 10.3390/ijgi11050316

Durdyev, S., Ashour, M., Connelly, S., & Mahdiyar, A. (2022). Barriers to the implementation of Building Information Modelling (BIM) for facility management. *Journal of Building Engineering, 46*, 103736. Advance online publication. DOI: 10.1016/j.jobe.2021.103736

Fan, C., Hou, J., Li, X., Song, G., Yang, Y., Liang, X., Zhou, Q., Imran, M., Chen, G., Wang, Z., & Lu, P. (2025). Efficient urban flood control and drainage management framework based on digital twin technology and optimization scheduling algorithm. *Water Research, 282*, 123711. DOI: 10.1016/j.watres.2025.123711 PMID: 40319783

Fang, X., Li, J., & Ma, Q. (2023). Integrating green infrastructure, ecosystem services and nature-based solutions for urban sustainability: A comprehensive literature review. *Sustainable Cities and Society, 98*, 104843. DOI: 10.1016/j.scs.2023.104843

Gargiulo, I., Garcia, X., Benages-Albert, M., Martinez, J., Pfeffer, K., & Vall-Casas, P. (2020). Women's safety perception assessment in an urban stream corridor: Developing a safety map based on qualitative GIS. *Landscape and Urban Planning, 198*, 103779. Advance online publication. DOI: 10.1016/j.landurbplan.2020.103779

Gholami, M., Torreggiani, D., Tassinari, P., & Barbaresi, A. (2022). Developing a 3D City Digital Twin: Enhancing Walkability through a Green Pedestrian Network (GPN) in the City of Imola, Italy. *Land (Basel), 11*(11), 1917. Advance online publication. DOI: 10.3390/land11111917

Green, D., O'Donnell, E., Johnson, M., Slater, L., Thorne, C., Zheng, S., Stirling, R., Chan, F. K. S., Li, L., & Boothroyd, R. J. (2021). Green infrastructure: The future of urban flood risk management? *WIREs. Water, 8*(6), e1560. DOI: 10.1002/wat2.1560

Huang, S., Tang, L., Hupy, J. P., Wang, Y., & Shao, G. (2021). A commentary review on the use of normalized difference vegetation index (NDVI) in the era of popular remote sensing. In *Journal of Forestry Research* (Vol. 32, Issue 1). DOI: 10.1007/s11676-020-01155-1

Islam, M. S., Yeasmin, T., Karmaker, S., Hossain, M. S., & Shi, L. (2023). Vegetation Cover Change Analysis during 1989-2020 of Coastal Barguna District, Bangladesh Using Remote Sensing and GIS Technology. *International Review for Spatial Planning and Sustainable Development, 11*(2), 259–277. Advance online publication. DOI: 10.14246/irspsd.11.2_259

Kampczyk, A., & Dybeł, K. (2021). The fundamental approach of the digital twin application in railway turnouts with innovative monitoring of weather conditions. *Sensors (Basel), 21*(17), 5757. Advance online publication. DOI: 10.3390/s21175757 PMID: 34502648

Kashyap, B., & Kumar, R. (2021). Sensing Methodologies in Agriculture for Soil Moisture and Nutrient Monitoring. *IEEE Access : Practical Innovations, Open Solutions, 9*, 14095–14121. Advance online publication. DOI: 10.1109/ACCESS.2021.3052478

Kumar, K., Labetski, A., Ohori, K. A., Ledoux, H., & Stoter, J. (2019a). Harmonising the OGC standards for the built environment: A CityGML extension for Landinfra. *ISPRS International Journal of Geo-Information, 8*(6), 246. Advance online publication. DOI: 10.3390/ijgi8060246

Kumar, K., Labetski, A., Ohori, K. A., Ledoux, H., & Stoter, J. (2019b). The LandInfra standard and its role in solving the BIM-GIS quagmire. *Open Geospatial Data. Software and Standards, 4*(1), 5. Advance online publication. DOI: 10.1186/s40965-019-0065-z

Kutzner, T., Chaturvedi, K., & Kolbe, T. H. (2020). CityGML 3.0: New Functions Open Up New Applications. *PFG -. Journal of Photogrammetry, Remote Sensing and Geoinformation Science, 88*(1), 43–61. Advance online publication. DOI: 10.1007/s41064-020-00095-z

Lee, D., & Lee, S. (2021). Digital twin for supply chain coordination in modular construction. *Applied Sciences (Basel, Switzerland), 11*(13), 5909. Advance online publication. DOI: 10.3390/app11135909

Liu, L., Li, B., Zlatanova, S., & van Oosterom, P. (2021). Indoor navigation supported by the Industry Foundation Classes (IFC): A survey. *Automation in Construction, 121*, 103436. DOI: 10.1016/j.autcon.2020.103436

Lopes, H. S., Vidal, D. G., Cherif, N., Silva, L., & Remoaldo, P. C. (2025). Green infrastructure and its influence on urban heat island, heat risk, and air pollution: A case study of Porto (Portugal). *Journal of Environmental Management, 376*, 124446. DOI: 10.1016/j.jenvman.2025.124446 PMID: 39921953

Madeira, R. N., Santos, P. A., Java, O., Priebe, T., Graça, E., Sarközi, E., Asprion, B., & Gomez, R. P. B. (2022). Towards Digital Twins for Multi-Sensor Land and Plant Monitoring. *Procedia Computer Science, 210*(C), 45–52. Advance online publication. DOI: 10.1016/j.procs.2022.10.118

Meschini, S., Pellegrini, L., Locatelli, M., Accardo, D., Tagliabue, L. C., Di Giuda, G. M., & Avena, M. (2022). Toward cognitive digital twins using a BIM-GIS asset management system for a diffused university. *Frontiers in Built Environment, 8*, 959475. Advance online publication. DOI: 10.3389/fbuil.2022.959475

Mourato, S., Fernandez, P., Pereira, L. G., & Moreira, M. (2023). Assessing Vulnerability in Flood Prone Areas Using Analytic Hierarchy Process—Group Decision Making and Geographic Information System: A Case Study in Portugal. *Applied Sciences (Basel, Switzerland), 13*(8), 4915. Advance online publication. DOI: 10.3390/app13084915

Mujib, M. A., Apriyanto, B., Kurnianto, F. A., Ikhsan, F. A., Nurdin, E. A., Pangastuti, E. I., & Astutik, S. (2021). Assessment of Flood Hazard Mapping Based on Analytical Hierarchy Process (AHP) and GIS: Application in Kencong District, Jember Regency, Indonesia. *Geosfera Indonesia*, *6*(3), 353. Advance online publication. DOI: 10.19184/geosi.v6i3.21668

Okonta, E. D., Vukovic, V., & Hayat, E. (2024). Prospective Directions in the Computer Systems Industry Foundation Classes (IFC) for Shaping Data Exchange in the Sustainability and Resilience of Cities. *Electronics 2024, Vol. 13, Page 2297, 13*(12), 2297. DOI: 10.3390/electronics13122297

Olanrewaju, O. I., Kineber, A. F., Chileshe, N., & Edwards, D. J. (2022). Modelling the relationship between Building Information Modelling (BIM) implementation barriers, usage and awareness on building project lifecycle. *Building and Environment*, *207*, 108556. Advance online publication. DOI: 10.1016/j.buildenv.2021.108556

Önden, İ., Pamucar, D., Deveci, M., As, Y., Birol, B., & Şüheda Yıldız, F. (2024). Prioritization of transfer centers using GIS and fuzzy Dombi Bonferroni weighted Assessment (DOBAS) model. *Expert Systems with Applications*, *238*, 121827. Advance online publication. DOI: 10.1016/j.eswa.2023.121827

Open Geospatial Consortium. (2017). *InfraGML*. LandInfra Core - Encoding Standard.

Ozel, B., & Petrovic, M. (2024). Green Urban Scenarios: A Framework for Digital Twin Representation and Simulation for Urban Forests and Their Impact Analysis. *Arboriculture & Urban Forestry*, *50*(1), jauf.2023.029. Advance online publication. DOI: 10.48044/jauf.2023.029

Popoola, O. A. M., Carruthers, D., Lad, C., Bright, V. B., Mead, M. I., Stettler, M. E. J., Saffell, J. R., & Jones, R. L. (2018). Use of networks of low cost air quality sensors to quantify air quality in urban settings. *Atmospheric Environment*, *194*, 58–70. Advance online publication. DOI: 10.1016/j.atmosenv.2018.09.030

Rasheed, M. W., Tang, J., Sarwar, A., Shah, S., Saddique, N., Khan, M. U., Imran Khan, M., Nawaz, S., Shamshiri, R. R., Aziz, M., & Sultan, M. (2022). Soil Moisture Measuring Techniques and Factors Affecting the Moisture Dynamics: A Comprehensive Review. In *Sustainability (Switzerland)* (Vol. 14, Issue 18). DOI: 10.3390/su141811538

Rodríguez-Espinosa, V. M., Aguilera-Benavente, F., & Gómez-Delgado, M. (2020). Green infrastructure design using GIS and spatial analysis: A proposal for the Henares Corridor (Madrid-Guadalajara, Spain). *Landscape Research*, *45*(1), 26–43. Advance online publication. DOI: 10.1080/01426397.2019.1569221

Senhorelo, A. P., Sousa, E. F., de, , Santos, A. R., dos, , Ferrari, J. L., Peluzio, J. B. E., & Carvalho, R. (2024). Application of Path Analysis and Remote Sensing to Assess the Interrelationships between Meteorological Variables and Vegetation Indices in the State of Espírito Santo, Southeastern Brazil. *Diversity*, *16*(2), 90. Advance online publication. DOI: 10.3390/d16020090

Tan, T., Mills, G., Papadonikolaki, E., & Liu, Z. (2021). Combining multi-criteria decision making (MCDM) methods with building information modelling (BIM): A review. *Automation in Construction*, *121*, 103451. DOI: 10.1016/j.autcon.2020.103451

Tan, Y., Liang, Y., & Zhu, J. (2023). CityGML in the Integration of BIM and the GIS: Challenges and Opportunities. In *Buildings* (Vol. 13, Issue 7). DOI: 10.3390/buildings13071758

Thakur, J. K., Singh, S. K., & Ekanthalu, V. S. (2017). Integrating remote sensing, geographic information systems and global positioning system techniques with hydrological modeling. In *Applied Water Science* (Vol. 7, Issue 4). DOI: 10.1007/s13201-016-0384-5

Todeschi, V., Pappalardo, S. E., Zanetti, C., Peroni, F., & De Marchi, M. (2022). Climate Justice in the City: Mapping Heat-Related Risk for Climate Change Mitigation of the Urban and Peri-Urban Area of Padua (Italy). *ISPRS International Journal of Geo-Information*, *11*(9), 490. Advance online publication. DOI: 10.3390/ijgi11090490

Tripathi, I., Froese, T. M., & Mallory-Hill, S. (2023). Applicability of BIM-IoT-GIS integrated digital twins for post occupancy evaluations. *Frontiers in Built Environment*, *9*, 1103743. Advance online publication. DOI: 10.3389/fbuil.2023.1103743

Tunio, I. A., Memon, S. A., Abbasi, H., Abro, A. A., Kumar, L., Tunio, M. A., Tunio, A. R., & Kumar, J. (2025). Spatiotemporal Assessment of Land Use Land Cover Change Detection and Normalized Difference Vegetation Index in Desert Area. *Rangeland Ecology and Management*, *101*, 126–139. DOI: 10.1016/j.rama.2025.03.008

Vignali, V., Acerra, E. M., Lantieri, C., Di Vincenzo, F., Piacentini, G., & Pancaldi, S. (2021). Building information Modelling (BIM) application for an existing road infrastructure. *Automation in Construction*, *128*, 103752. Advance online publication. DOI: 10.1016/j.autcon.2021.103752

Wang, Y., Wang, Z., Zhao, D., Yu, X., Cheng, T., Bao, G., & Wang, Z. L. (2021). Flow and level sensing by waveform coupled liquid-solid contact-electrification. *Materials Today Physics*, *18*, 100372. Advance online publication. DOI: 10.1016/j.mtphys.2021.100372

Xia, H., Liu, Z., Efremochkina, M., Liu, X., & Lin, C. (2022). Study on city digital twin technologies for sustainable smart city design: A review and bibliometric analysis of geographic information system and building information modeling integration. *Sustainable Cities and Society*, *84*, 104009. DOI: 10.1016/j.scs.2022.104009

Yan, S., & Tang, J. (2021). Optimization of green space planning to improve ecosystem services efficiency: The case of chongqing urban areas. *International Journal of Environmental Research and Public Health*, *18*(16), 8441. Advance online publication. DOI: 10.3390/ijerph18168441 PMID: 34444194

Zhang, H., & Srinivasan, R. (2020). A systematic review of air quality sensors, guidelines, and measurement studies for indoor air quality management. In *Sustainability (Switzerland)* (Vol. 12, Issue 21). DOI: 10.3390/su12219045

Zhao, T., Song, C., Yu, J., Xing, L., Xu, F., Li, W., & Wang, Z. (2025). Leveraging Immersive Digital Twins and AI-Driven Decision Support Systems for Sustainable Water Reserves Management: A Conceptual Framework. *Sustainability 2025, Vol. 17, Page 3754, 17*(8), 3754. DOI: 10.3390/su17083754

Zhu, J., & Wu, P. (2021). Towards effective bim/gis data integration for smart city by integrating computer graphics technique. *Remote Sensing (Basel)*, *13*(10), 1889. Advance online publication. DOI: 10.3390/rs13101889

Chapter 4
Leveraging City Information Modeling (CIM) and GIS for Sustainable Road Infrastructure:
The SIC Housing Access Road in Mfou, Cameroon

Cyntia Kengne Ngonda Tassi
National Advanced School of Publics Works, Yaoundé, Cameroon

Bikie Gerald Anicet
https://orcid.org/0009-0003-2068-3494
National Advanced School of Publics Works, Yaoundé, Cameroon

Roger Blaise Limaleba
National Advanced School of Publics Works, Yaoundé, Cameroon

Yassine Mouniane
https://orcid.org/0000-0002-6037-5301
Faculty of Sciences, Ibn Tofail University, Kenitra, Morocco

ABSTRACT

This chapter presents a digitally integrated methodology for the planning, design, and cost estimation of a critical road infrastructure project in Mfou, Cameroon, supporting a major social housing initiative by the Société Immobilière du Cameroun

DOI: 10.4018/979-8-3373-4058-6.ch004

(SIC). The chapter details the use of Geographic Information Systems (GIS) and City Information Modeling (CIM) to develop a digital twin of the access road, incorporating topography, utility networks, road geometry, and planning constraints. Through the combined use of Autodesk InfraWorks, AutoCAD Civil 3D, and Primavera P6, the project is modeled in 3D, scheduled in 4D, and financially assessed in 5D, with costs estimated at 408,361,987 FCFA. The chapter highlights the benefits of BIM–GIS integration, including enhanced data accuracy, improved interprofessional collaboration, real-time stakeholder engagement, and significant time and cost efficiencies compared to traditional methods. It proposes a scalable approach to digital transformation in infrastructure development for emerging regions.

INTRODUCTION

Urbanization has emerged as a key global trend of the 21st century. As forecasted by the United Nations, over 66% of the world's population will live in urban areas by 2050, with Africa growing at one of the fastest rates. Rapid urbanization creates considerable pressure on cities to provide good infrastructure, sustainable housing, and efficient transport systems. Roads, in particular, are a critical element of urban connectivity, providing access to economic, social, and administrative functions (Shamaee et al., 2024; Sharma et al., 2025).

In an ever-changing world where cities are becoming key centres of economic, social and cultural development, urban planning and infrastructure management are becoming crucial issues (Nijkamp et al., 2024). According to the United Nations, more than 66% of the world's population is expected to live in urban areas by 2050 (Norrman, 2023). However, having well-maintained road networks that provide the level of service expected by users has now become a key factor in development strategies for Africa, and Cameroon in particular (Nchang et al., 2024). 'Managing urban areas has become one of the most important development challenges of the 21st century', said the Director of the Population Division, John Wilmoth, at a press conference at UN headquarters in New York. At the heart of this statement concerning the management of urban areas is the construction and efficient management of road infrastructure, which is the foundation of urban connectivity and citizen mobility.

Recently, the way we model urban information is being taken to the next level in Architecture, Engineering and Construction. Furthermore, City Information Modeling (CIM) is expanding in popularity and usage not only in civil engineering, but mechanical and electrical engineering too (Alizadeh et al., 2016; Šimenić, 2021; Najafi et al., 2023). CIM does not just include threedimensional (3D) models, it can incorporate other dimensions such as 4D planning, 5D costing and many more. Specifically within the commune of Mfou, Cameroon, construction and management of road infrastructure has many issues ranging from some neighbourhoods being uninformed, absence of urban planning and above all poor maintenance (Gil et al., 2011; Beil and Kolbe, 2017; Souza and Bueno, 2022; Rezvani et al., 2023). The access road to the Real Estate Company of Cameroon at EKOKO II, which remains undeveloped and without any network, has strategic importance within the urban fabric of this community and is essential within the context of access for social housing and local economic development (Xu et al., 2014; Ait-Lamallam et al., 2019; Deprêtre et al., 2023).

In Cameroon as in many developing countries, the road infrastructure is a springboard for local and national development. Good road infrastructure allows for improved mobility, economic growth, and access to social services. Despite this, all municipalities face some combination of poor planning, poor maintenance, and absence of integrating data across construction phases. Many studies have examined the enhancement of road infrastructure construction. Although CIM is becoming more popular in developed countries, its use across the urban context in Africa is limited. In Cameroon, urban development schemes deploy the standard 2D design method with no integration of design, cost estimation and construction scheduling to deliver projects. This fragmented approach results in inefficiencies, delays, and cost overruns, further diminishing the opportunities for economic development and accessibility for Cameroonian urban residents. Furthermore, while various studies have examined digital modeling for building projects, few studies have examined its use in road infrastructure. Of course, the roads are a vital aspect of accessibility and economic development (Gil et al., 2011; Beil & Kolbe, 2017; Rezvani et al., 2023). It is imperative to show that CIM can be used to design, plan, and manage urban roads across all project dimensions (3D design, 4D time, and estimate construction costs within the collaborative digital platform). There are no case studies that use CIM in a municipality in Cameroon, which is a significant gap in the body of research. This study aims to provide a case study of CIM for an access road to the Real Estate Company's (SIC) housing at Ekoko II in the Mfou commune.

While Building Information Modeling (BIM) considers primarily individual buildings, City Information Modeling (CIM) progresses those practices to the scale of urban environments, bringing together infrastructure, topography, and utilities into one approach for a comprehensive urban planning and management method. However, most of these focus on specific technical aspects, such as pavement design or site management; few studies really look at the use of new technologies, such as CIM, to improve the entire road infrastructure construction process. Indeed, a better understanding of the application of CIM can make a significant contribution to improving the planning, design, construction and management of urban roads, which could have a positive impact on the quality of life of citizens and the sustainable development of cities (Dantas et al., 2019; Castañeda et al., 2024). So how can

CIM contribute to the construction of sustainable road infrastructures to guarantee more effective results and improve decision-making?

The main objective of this study is therefore to develop a detailed (3D) CIM model of the SIC housing access road at Ekoko II, integrating all the geometric and functional elements, to serve as a reference for the subsequent phases of the project.

To achieve our goal, we will have the following specific objectives:

- Design the 3D model of the access road,
- Draw up the schedule and estimate the costs of the project,
- Manage a CIM collaboration platform for monitoring the project.

This work will be divided into three chapters. The first will provide a general overview of road infrastructure and CIM, laying the foundations for our study. The second will set out the study area and the methodology used to carry out our research, leading on to the final chapter devoted to the presentation and analysis of our results.

LITERATURE REVIEW

The development of urban infrastructure has progressed from traditional construction management methods toward data-driven, integrated approaches that integrate spatial, temporal, and financial aspects. The change from heritage 2D design tools to multidimensional information models has disrupted the ways cities plan, build and maintain public works. Among these innovations are Building Information Modeling (BIM) and its urban-scale extension, City Information Modeling (CIM), which has emerged as important enablers of smart and sustainable urban development (Lasorella & Cantatore, 2025). This background section provides an overview of the conceptual evolution of CIM, its relationship to BIM, its technological underpinnings, and its application in road infrastructure and sustainable urban management, especially in developing countries such as Cameroon.

Evolution from BIM to CIM

In the late 1990s, Building Information Modeling (BIM) emerged as a new digital process for the building industry that provided a shared digital representation of a building's physical and functional features (Berlato et al., 2025). BIM systems provided better workflow productivity by incorporating geometry, materials, schedule, and costs within a single digital environment to better coordinate architects, engineers, and contractors. However, as cities grew, the role of BIM became limited for more complex city projects where there was a demand to service large-scale heterogeneous networks of infrastructure beyond buildings. City Information Modeling (CIM) emerged to fill this gap. CIM takes the idea of BIM — Building Information Modeling — to an entire city system. It incorporates data from multiple sources such as the physical environments surrounding cities including topography, road networks, utilities, and environmental (weather) data into the same model. Another important characteristic of CIM is that it can represent urban structures in both the vertical and territorial dimensions (Yu et al., 2025). BIM primarily deals with vertical construction, but CIM integrates the horizontal aspect of urban construction by considering roads, bridges, drainage systems, vegetation, or built environments into 3D city models or sometimes even 5D simulations. Integrating all this information permits CIM to assist urban planners and decision makers, during the planning and design process, to test scenarios, deliver spatial analytics, and design infrastructure for sustainability. Essentially, BIM is object centered (relates to building elements) while CIM is context centered (the interaction of urban elements). Again, it is not just scale that differentiates them, it is the intention. BIM emphasizes efficiency in the design and construction process and CIM emphasizes urban governance, resilience, and sustainability.

Conceptual Framework of CIM

CIM operates at the intersection of geospatial information systems (GIS), BIM, and Internet of Things (IoT) technologies. It harnesses geospatial positioning, digital twins, and data interoperability to describe cities as multi-dimensional, continuous, and evolving systems, with the underlying conceptual framework of CIM built on three pillars: (1) Integration: bringing together spatial, semantic, and temporal data from disparate sources; (2) Visualization of multi-scale 3D/4D/5D simulations; and (3) Collaboration between stakeholders through shared digital platforms (e.g. Autodesk Construction Cloud, Bentley OpenCities). New developments in semantic modeling languages like CityGML and IFC (Industry Foundation Classes) have standardized the exchange of data between building information modeling (BIM) and GIS environments. These standards facilitate the representation of data through a top-down hierarchy from terrain to buildings, providing consistency across scales. CIM also integrates sensor data from IoT devices that can be utilized to monitor conditions of roads, traffic, and energy consumption. This allows predictive maintenance and optimizes infrastructure performance. In addition, CIM is aligned with the digital twin concept, a virtual version of real-world physical assets that continuously updates based on real time information. These updates provide the basis for scenario-based simulations, such as the impact of rainfall events on flooding in deep urban drainage systems, or estimating traffic flows by regional emissions.

Such analytical processes are necessary for evidence-based urban policy and all associated Sustainable Development Goals (SDG 9 and SDG 11).

Applications of CIM in Road Infrastructure

The road infrastructure is an essential enabler of urban mobility, economic development, and spatial equity. Nonetheless, conventional road design is often characterized by disparate workflows, limited collaboration, and poor communication among multiple stakeholders. CIM is a proven solution to circumvent these issues by incorporating an integrated workflow through staging any project–planning, design, construction, and operation (Das, 2024).

During the planning phase, CIM is useful in visualizing site conditions and land use compatibility using georeferenced data. Prior to any physical construction starting, professionals are able to examine several design alternatives that consider above all slope, the integrity of the soil in addition to analyzing potential ecological risk. For the design and construction process, CIM can provide engineers with intricate models that include a 3D format that can provide coordination of drainage, lighting, fibre optics, and all other subsurface utilities in one location. Adding time (4D) and cost (5D) variables can allow better scheduling of construction and more accurate estimating of project budgets (Taheri & Sobanjo, 2024). CIM is intended as a decision-support system during the post-construction process, mainly with regard to scheduling maintenance, evaluating assets, and monitoring environmental impact. Using CIM for road management for example will provide automated monitoring patterns of deterioration and assessing risk in order to prioritize improvements, thus limiting both costs and the impact of disturbance. This could be particularly helpful for countries such as Cameroon where municipalities may only rely on their own resources (people) to monitor infrastructure, and do not maintain any type of database to store infrastructure asset data. Furthermore, this work aligns with the world wide need for more sustainable mobility related to road projects. In particular, CIM allows users to model the interaction of network with other environmental variables (e.g., air quality, runoff, vegetation cover) for a better understanding of how sustainable patterns of ecological footprints and resource (e.g., materials, etc.) usage can be minimized.

CIM for Urban Governance and Smart Cities

CIM functions not just as a technical aid, but as a tool for understanding governance. Through CIM, local authorities can gain insight into city dynamics, simulate policy scenarios, and coordinate investments. In smart cities, CIM is increasingly utilized to manage services, traffic, and energy systems in a common spatial context.

With the integration of artificial intelligence and IoT, real-time data-based public governance can be facilitated through CIM. A primary challenge to urban governance in Africa is the fragmentation of spatial and administrative data. Cities often have separate municipal departments that use outdated maps or unrelated databases. CIM is capable of combining several of these data sets into one interoperable model that allows for the sharing of information with your stakeholder community. Linking CIM to cadaster and demography data could provide a more dependable basis for local planning decisions around road extensions or even social housing, to cite two examples. However, CIM does more than manage infrastructure; The CIM system facilitates community engagement. Your stakeholders, along with residents and the general public, could have access to web-based visualization interfaces and be able to engage with, propose changes to, or simply follow along with the construction and Digital Twin of the city. Participatory CIM systems support the development of public trust and accountability, vital components of sustainable urban governance.

CIM Implementation Challenges

While CIM has positive aspects, it has certain barriers for implementation, especially in developing contexts. The first is technical challenges, such as data interoperability, not having a national standard for geospatial data, and limited computing infrastructure. When combining data — CAD, GIS, photogrammetry, etc. — typically these require a more structured series of conversion workflows and the redoing of quality control. Second, institutional barriers relate to the absence of single responsibilities between central and local jurisdictions, and private contractors. In order to implement CIM successfully, there must be a framework for coordination between departments and a governance framework to address issues of data ownership and source.

Thirdly, the building capacity constraints is an important contributing factor. The move to CIM calls for engineers, urban planners, and policymakers to be well trained in 3D modelling, GIS, and project management software. As CIM principles have not yet been fully embedded in capacity-building programs or the university curriculum in Cameroon and other African countries, progress is being made in specific universities and professional organizations. Finally, the financial constraints are significant. Software licenses, data acquisition costs (i.e. LiDAR, satellite imagery), and server infrastructure all come with high start-up costs. However, it has been suggested in the studies that there are tangible benefits in the long run and that this may compensate for some of the initial costs in the form of reduced design mistakes, shortened timeframes for projects, and more efficient maintenance.

Integration of CIM with Emerging Technologies

The effectiveness of CIM is improved through the integration of new digital technologies. For example, Artificial Intelligence (AI) could be employed to automate object recognition in LiDAR point clouds or use an anomaly detection approach to detecting anomalies in a road network. Machine learning algorithms could be used to predict traffic congestion or material degradation by referring to historical information in the realm of predictive maintenance. Similarly, the use of Augmented and virtual reality applications is enabling engineers and decision-makers to visualize projects in a richer and more immersive manner. An example is the ability to "walk-through" proposed road alignment virtually prior to construction! Not only does this improve understanding of the design, but also it provides an opportunity for public engagement. In addition, blockchain technology can augment CIM by maintaining data integrity and traceability for construction workflow including design approvals and payment certificates. This type of integration enhances accountability for public infrastructure projects. Cloud computing is also favorable to collaborative environments that are conducive to Construction Information Modeling (CIM), where several stakeholders can work on the same model in real time. The Autodesk Construction Cloud, for example, which was utilized in this research, provides safe cloud storage to manage documents, role-based access, and version control—all important considerations for complex urban projects.

CIM and Sustainable Development in African Contexts

The rapid urbanization rate in Africa, which often exceeds 4% a year, presents both challenges and opportunities for the use of CIM. Nigeria's poorly coordinated spatial expansion, informal settlements and constrained maintenance budgets will all contribute to urban inefficiencies. CIM can potentially be used as a strategy to foster more resilient and equitable urban development by accounting for environmental, social or economic parameters during the designing of the infrastructure. For example, in Cameroon, cities such as Yaoundé and Douala may have intermittent problems with road deterioration, flooding, and a lack spatial plan. When CIM is applied to road projects like the access road to housing SIC in Mfou, planners will know to anticipate drainage problems, design slopes better, and reduce waste materials. CIM provides a more robust analysis of topography, hydrology, and utilities that support decision making that the correct projects are developed on the basis of envirosocial projects concurrently with development. CIM promotes learnings at the institutional level as well as developing physical infrastructure. Once models are developed from several studies, local municipalities can produce templates and best practice for future work much quicker. Knowledge transfer is a key aspect to develop local expertise and lessen the reliance on external consultants.

Future Perspectives

In the future, the advancement of CIM will become increasingly connected with sustainable digital governance. The next generation of CIM systems will provide a fusion of urban analytics, artificial intelligence (AI), and real-time sensors to develop self-updating urban twins that will enable climate adaptation, disaster risk reduction, and circular economy. In the field of road infrastructure, CIM should be positioned to have a key role in improving materials and energy efficiency. Engineers can simulate the life-cycle performance of a chosen design and select the design that best reduced carbon emissions and operational costs. CIM will be an integral part of sustainable infrastructure development with principles of the practice of green construction embodied in low-impact materials and intelligent drainage. The democratization of open-source CIM platforms might allow smaller municipalities to access models used for large municipalities. Schools of thought among academia, government, and industry should be connected and leveraged to improve urban growth that is inclusive and sustainable.

METHODOLOGY

Presentation of the Site

Mfou is a municipality in Cameroon, located in the Central Cameroon region, about 30 km from Yaoundé, capital of the Mefou and Afamba department, with coordinates 4°27'0''N and 11°38'00''E. It has about 15 localities, including EKOKO II, which is the site of the road section of our study project. This is the access road to the 10,000 housing units of the SIC (Figure 1). The Mfou commune, located 30 km from Yaoundé (4°27′N, 11°38′E), lies on the Mbélé plateau with altitudes between 600–800 m. Its tropical monsoon climate (Am, Köppen–Geiger) brings an average annual rainfall of 1,686 mm and temperature of 23.5 °C. The Nyong River and its tributaries shape the local hilly landscape.

It has a hilly relief characteristic of the Mbélé plateau. The minimum altitude is around 600 m in the valley of the River Nyong, while the maximum altitude is around 800 m at the top of the hills surrounding the town. The terrain is characterised by a succession of small hills of volcanic origin and valleys, often marshy and crossed by rivers.

The weather in this region is tropical. The town of Mfou is characterised by many months of heavy rainfall. There is only a short dry season, and this is not very marked. According to the Köppen-Geiger classification, the climate prevailing in this region is classified as Am (hot monsoon climate). The average annual temperature recorded in Mfou is 23.5°C. Annual rainfall is 1,686 mm, according to meteorological records. The summer months are June, July, August, September

The town of Mfou benefits from a dense and permanent hydrographic network, made up of rivers, streams and springs. The River Nyong, one of Cameroon's major rivers, flows through the town into the Gulf of Guinea. Other notable rivers include the Mfoundi, Nyong Ekelle and Lokoué. The town is located on the P11 provincial road, 26 km south-east of the capital Yaoundé (Place Ahmadou Ahidjo).

According to the 2005 census, the commune of Mfou had a population of 37,209. The town of Mfou itself had a population of 10,533, representing around 28% of the commune's total population.

Figure 1. Situation map of the study area

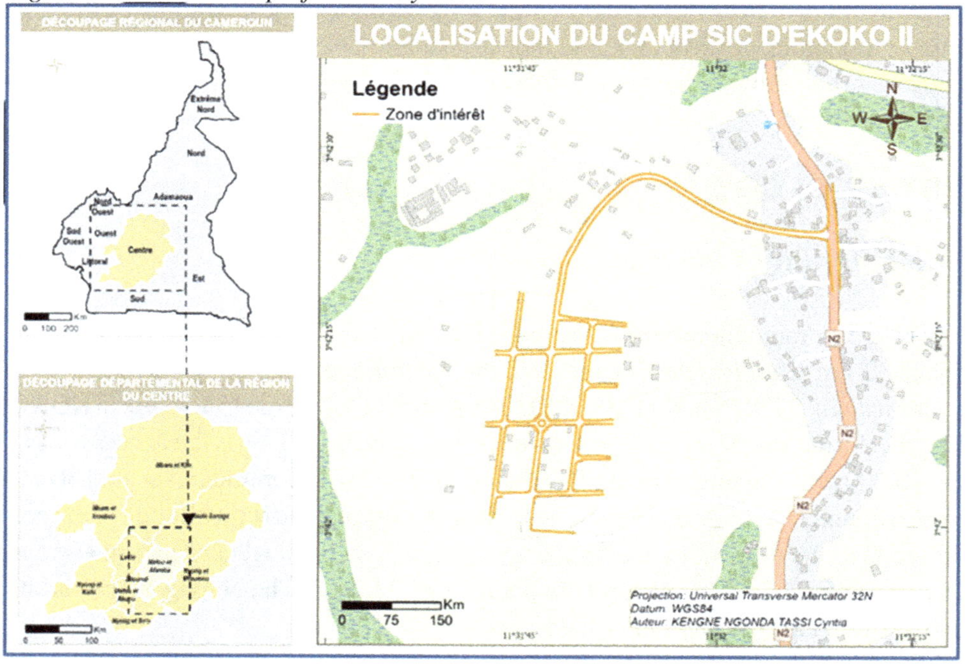

Data

A formal request was submitted to PIZZAROTTI BRANCH CAMEROUN for specific information. A favourable response to this request was the crucial first step in our data collection. The data is varied and includes

- A site survey plan and precise topographical survey of the access road to the study site, enabling us to understand the current site conditions and plan future works.
- Drinking water supply plan: providing essential information on existing infrastructure and future needs.
- The electricity network distribution plan, including the position of lines, transformer stations and connections to dwellings, and power balances.
- The fibre optic distribution plan, to identify connectivity requirements for future developments.
- The BIM model social housing, providing a detailed representation of residential structures.

Approach

Data processing takes place in four stages, including 3D modeling, project activity planning (4D), cost estimation (5D) and management of the collaboration platform. Our project flowchart is shown in Figure 2.

Figure 2. Organisation chart of the CIM project

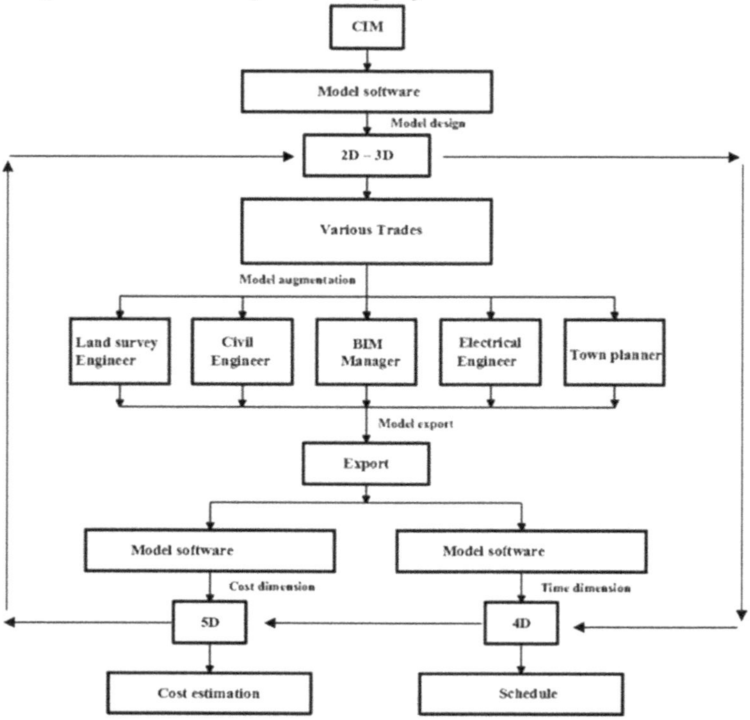

2D Modeling of the Road Project

2D modeling of road projects provides an accurate representation of the terrain and is essential for the planning and construction of road infrastructure. It provides a common basis for engineers, architects and other stakeholders to better understand the site and plan effectively.

This modeling begins with the use of topographic survey data in Covadis software to obtain the characteristics of the terrain, including altitude, slopes and natural features, followed by the removal of construction elements to calculate the Digital Terrain Model (DTM), and then the geometric analysis of the road, which involves:

- Plan layout
- Longitudinal profile
- Cross-section

These stages are followed by the production of various calculation summary plans, in particular:

- Project calculation reports
- The volumes and surfaces of the materials used
- Quantities of materials

Integrated 3D–5D Modeling and Collaborative Platform

Within the framework of City Information Modeling (CIM), 3D, 4D, and 5D modeling work together to facilitate the creation of a digital twin of infrastructure projects, representing geometry, scheduling, and cost information in one collaborative environment.

The project is grounded in a 3D model, which accurately represents the road. The 3D model represents the road's geometry, alignment, longitudinal and cross-sections, materials and quantities, underground utilities, and all facilities associated with the road. The 3D model was prepared in Autodesk Infraworks, and its function is not simply for visualization; designers are able to visualize how the road interacts with the existing environment and identify conflicts in design, such as slopes or complicated intersections, all enabling analyses to be integrated. The 3D model serves as the basis of all subsequent analyses.

While leveraging the 3D model, the 4D modeling phase incorporates the time dimension, linking construction activities back to the project schedule. After exporting the 3D data, including drawings and quantities, the different parts of the construction process are sequenced in chronological order to develop a dynamic project schedule. This data integration allows stakeholders to visualize the development of the works, plan for resource availability, and identify potential conflicts with scheduling. Construction planning was carried out using Primavera software, which has the capacity for 4D simulation and scheduling management.

During the 5D modeling phase, we introduce an economic or financial dimension by associating quantities and materials with cost indicators. Utilizing CIM tools, quantity takeoffs are automatically obtained from the models to provide an accurate and reliable cost estimating ability. This technique is more reliable than manual estimates as it is based on applications of two-dimensional drawings, which can significantly increase risks of going over budget. Also, cost analysis allows for the possible evaluation of different design alternatives which maximize either economic efficiency or material utilization.

To assist with coordination, all of these models are uploaded to a shared collaborative platform, the Autodesk Construction Cloud (ACC). The ACC platform offers engineers, architects, and project managers the ability to share models, documents, and updates simultaneously from a single source. The platform creates transparency with role-based access, as well as automated notifications to support shared communication and decision-making. The 3D, 4D, and 5D aspects in ACC allow model components to become complementary pieces of synchronized project management, which allows the project lifecycle, from design to execution, to be visible to all participants.In summary, this integrated CIM approach offers a robust digital environment that significantly improves project visualization, improves scheduling and cost control, and strengthens collaboration between stakeholders during the road infrastructure project development process.

RESULTS

Technical Analysis of the Road

This analysis provides results on the precise geometric characteristics of the road. It also assesses construction material requirements.

Plan Axis and Tabulations

It is essential that the plan axis designed respects all the parameters of the constraining points to avoid inhabitable areas as far as possible. The ARP standard is used, with a reference speed of 60 km/h. The designed plan axis is shown in Figure 3 below.

Figure 3. Plan axis

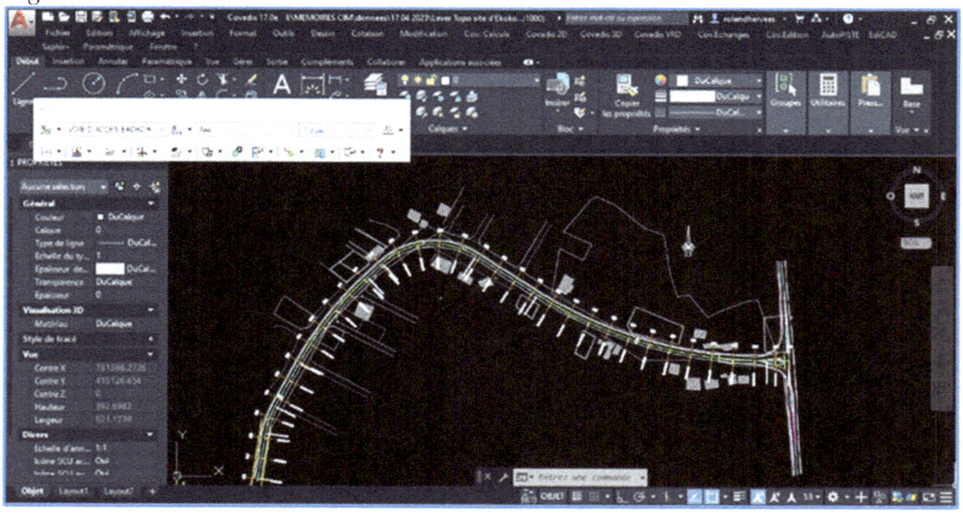

Longitudinal Profile

The alignment begins with that of the profile following the natural ground, then comes the design of the proposed line, known as the red line. The resulting longitudinal profile of the project is shown in Figure 4 below.

Figure 4. Longitudinal Profile

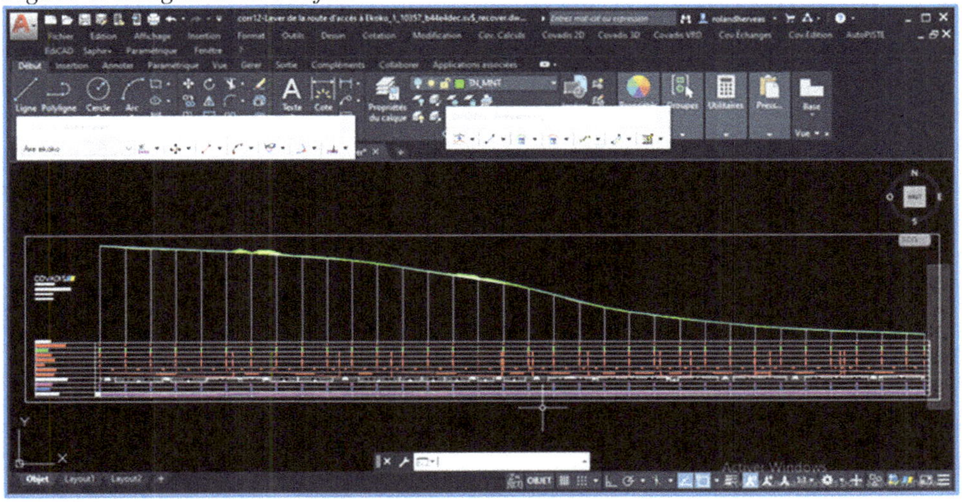

Cross-Section

The typical cross-section includes all materials and layers of the pavement; it can be drawn to the left or right of the centerline. The basic characteristics of the road project are as follows: a lateritic gravel sub-base layer, 25 cm thick; a crushed gravel base layer, 20 cm thick; a 5 cm thick bituminous concrete wearing course, and underground utilities.

Figure 5. Cross- section

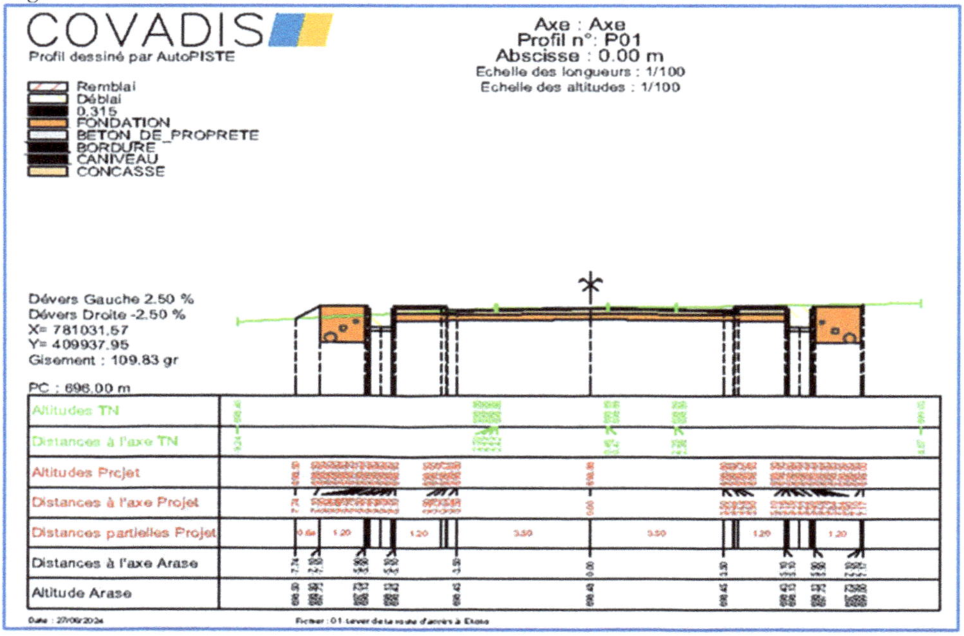

Generation of Earthworks and Layer Volumes

Based on the project calculations, we can automatically generate the quantities of cut and fill (Table 1) and the quantities of each material required for the project (Table 2).

Table 1. Earthworks

Num	Abscisse	Longueur	Surface_ Deblai	Surface_ Remblai	VolPart_ Deblai	VolPart_ Remblai	VolCum_ Deblai	VolCum_ Remblai
P01	0	12.5	8.5	0.12	106.2	1.5	106.2	1.5
P02	25	25	6.21	0.18	155.2	4.4	261.4	5.9
P03	50	25	6.83	0.07	170.3	1.8	431.6	7.7
P04	75	25	7.36	0.2	180.9	5.2	612.6	12.9
P05	100	25	9.27	0.05	231.6	1.3	844.2	14.2
P06	125	25	14.04	0	351	0	1195.3	14.2
P07	150	25	16.56	0	413.6	0	1608.8	14.2
P08	175	25	13.74	0	342.7	0	1951.6	14.2

continued on following page

Table 1. Continued

Num	Abscisse	Longueur	Surface_ Deblai	Surface_ Remblai	VolPart_ Deblai	VolPart_ Remblai	VolCum_ Deblai	VolCum_ Remblai
P09	200	25	13.36	0	334	0	2285.6	14.2
P10	225	25	14.9	0	372.4	0	2657.9	14.2
P11	250	25	14.27	0	357.2	0	3015.2	14.2
P12	275	25	12.28	0	307.3	0	3322.4	14.2
P13	300	25	11.35	0	284.5	0	3606.9	14.2
P14	325	25	10.11	0	255.2	0	3862.1	15
P15	350	25	12.2	0	304.4	0	4166.5	15
P16	375	25	13.41	0	335.4	1	4502	15
P17	400	25	11.37	0	281.7	0	4783.7	15
P18	425	25	8.23	0.05	205.1	1.3	4988.8	16.2
P19	450	25	8.52	0.01	212.9	0.2	5201.7	16.4
P20	475	25	6	0.09	150	2.4	5351.6	18.8
P21	500	25	5.31	0.2	132.7	5.1	5484.3	23.9
P22	525	25	5	0.23	124.7	5.8	5609	29.7
P23	550	25	4.21	0.27	104.9	6.7	5713.9	36.4
P24	575	25	5.52	0.15	138	3.6	5851.9	40.1
P25	600	25	7.29	0.03	182.7	0.7	6034.6	40.8
P26	625	25	8.97	0.02	224.8	0.5	6259.4	41.2
P27	650	25	6.39	0.09	159.4	2.3	6418.9	43.6
P28	675	25	8.58	0.09	213.6	2.3	6632.5	45.9
P29	700	25	8.34	0.1	207.8	2.5	6840.3	48.4
P30	725	25	10.38	0.01	258.8	0.2	7099	48.6

Table 2. Layer volumes

Num	Structure Materials	Quantity	Units	Unit Price (excl. tax)	Total Price (incl. tax)
GC-030	Supply and implementation of lean concrete	670,347	m³	0	0,00
GC-031	Supply and implementation of crushed material	1025,589	m³	0	0,00
GC-157	Supply and implementation of 0/315 aggregate	195,722	m³	0	0,00
GC-158	Supply and implementation of foundation	4048,481	m³	0	0,00
GC-162	Supply and implementation of gutter	50,072	m³	0	0,00
GC-163	Supply and implementation of curb	57,126	m³	0	0,00
Preparatory Works					

continued on following page

Table 2. Continued

Num	Structure Materials	Quantity	Units	Unit Price (excl. tax)	Total Price (incl. tax)
TP-001	Site installation	1,00		0,00	0,00
TP-002	Temporary site signage	1,00		0,00	0,00
TP-003	Setting out, phasing plan, and control	1,00		0,00	0,00

3D Model of the Project

Having designed our project in 2D, we can then visualise it in 3D and observe how it fits into the terrain using Infraworks software. This road modeling has facilitated the road alignment through the software's various commands, followed by the modelling of underground utilities according to the NF 98-332 standard, and then the insertion of ancillary features such as green spaces, vehicles, and vegetation.

Figure 6. 3D Model of the road

Figure 7. 3D model of the road and BIM buildings

Project planning (CIM 4D)

This model provides stakeholders with a clear understanding of the project schedule, outlining the implementation phases from initiation to handover. The project schedule was developed by configuring the EPS and OBS, and then breaking down the work into a WBS, as depicted in Figure 8.

Figure 8. Creation of project WBS

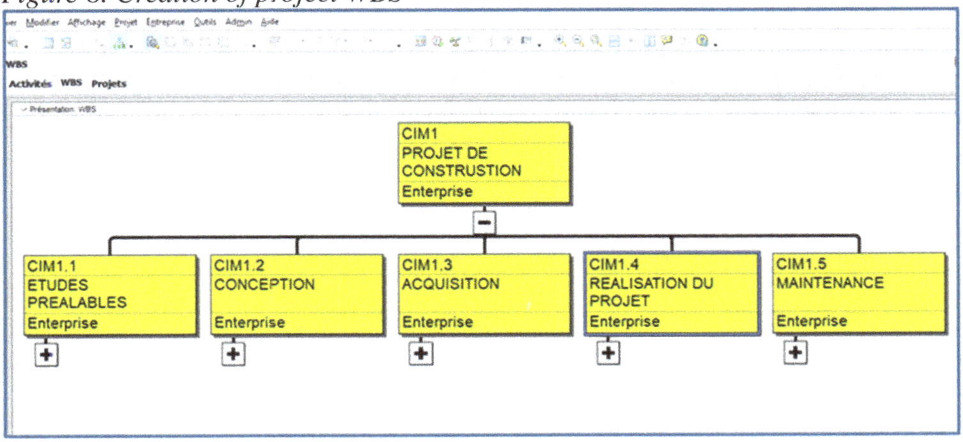

Project Cost Estimation

A 5D model was created by exporting quantity and material data. Costs were automatically calculated based on national market unit prices. For this project, we focused on material costs only. The budget was established by inputting an estimated total cost of 408,361,987 FCFA for the selected EPS node into the budget journal. This cost covers materials only and is shown in Figure 9.

Figure 9. Project budget

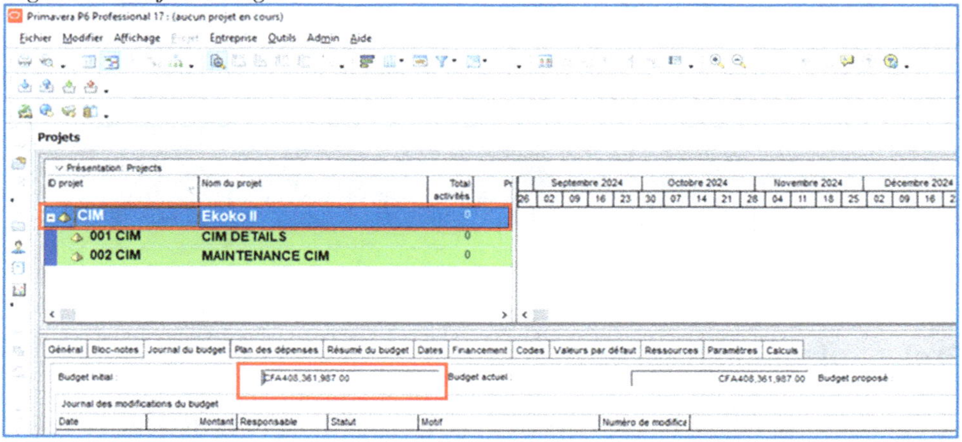

Collaboration plateform

Autodesk Construction Cloud (ACC) was utilized to facilitate the implementation of Building Information Modeling (BIM) on the project. The platform enabled the upload and sharing of project information and design data, such as AutoCAD drawings, Infraworks models, and various file formats, with all stakeholders. Following project creation in ACC, members were added and assigned specific access levels and roles. An automated invitation email was sent to each member, with the project creator assuming the role of administrator. As depicted in Figures 10 and 11, all individuals associated with a project are considered project members.

Figure 10. Project creation

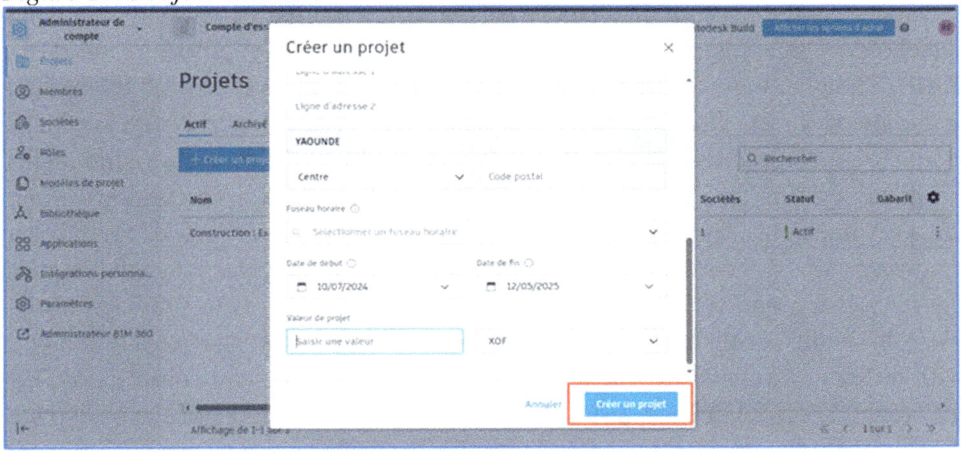

Figure 11. Define members roles and access

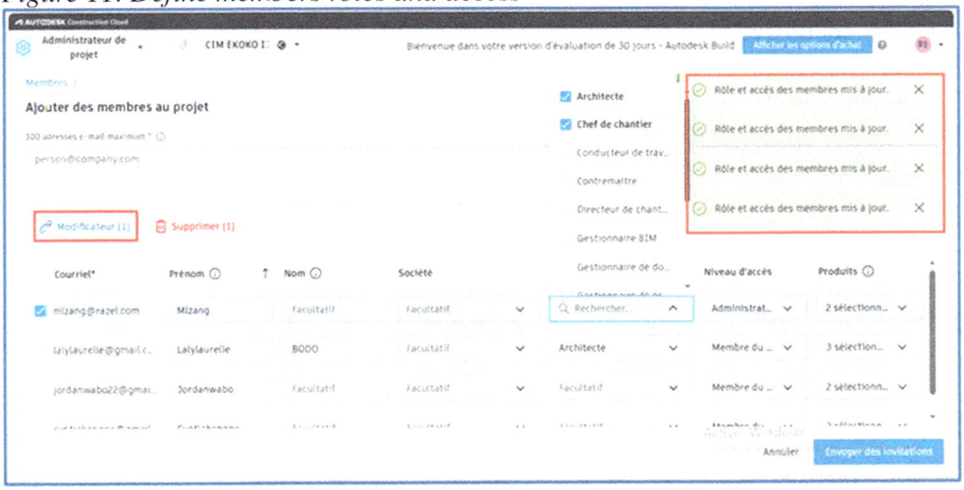

DISCUSSION

From the initial design stage, where we defined the road's geometric characteristics, to the 3D visualisation that enabled us to understand its integration into the environment, every step was crucial (Irschick et al., 2022; Takva et al., 2024).

Developing a detailed schedule and cost estimate ensured rigorous project management, while the use of a collaborative platform optimised coordination among stakeholders, guaranteeing overall project success (Deng et al., 2016; Akob et al., 2019). The plan view and the cross-sections provided a linear view of the project and allowed for 2D planning of the subsequent construction steps, seen in (Gikas et al., 2012; Khoshamadi et al., 2023). This step defined the necessary alignments for planning. The longitudinal profile illustrated the alteration in elevation along the road line. This visualization is critical for identifying sections of the project requiring excessive earthwork. The typical road section was created to standardise the road cross-sections. The purpose of the standardisation is to provide consistency in quality of the construction. The earthwork volume was generated along with different layers to provide necessary materials for the road construction. By developing a model builder, a substantial and detailed 3D model of the study area was developed; subsequently, this allowed for the visualization of the project in its environment. Visualization of the project in its environment provided a realistic and immersive view of the road's future condition, which included various utilities (Wang et al., 2019; Jovanović et al., 2020; Xu et al., 2021; Qi et al., 2024).

Creating a comprehensive project schedule organized the various stages of construction, identified relevant dependencies between different tasks, and established target dates. The project manager must effectively manage time to ensure the project is completed on time and within budget. Cost estimation was also very important to create a preliminary budget to manage costs and minimize overspend. This step allows the project to remain financially sustainable throughout the period of performance. The web platform supports communication and collaboration among project stakeholders such as engineers, architects, urban planners, managers, contractors, and possibly the local municipality. This platform supports project monitoring, enhanced decision-making, and appropriate change management in real time, helping to deliver less complicated built projects, organizations, Departments of Transportation, and municipalities as safely and sustainably as possible.

Increasing literature suggests that City Information Modeling (CIM) represents a cutting-edge framework for urban design and infrastructure management. According to Souza and Bueno (2022), "CIM is BIM applied to the territorial scale. Rather than simply providing cities with a digital space to leverage spatial, semantic, and temporal data, CIM enables public sector actors, engineers and planners to have "holistic" views of urban systems and respond to conditions in simulated scenarios." Instead of interpreting CIM solely as a technology, it should be seen as a decision-support system that can lead to more sustainable, efficient, and resilient cities. Their systematic examination reveals three key dimensions of CIM use: (1) urban planning and design, (2) infrastructure management, and (3) environmental assessment. In the urban planning domain, CIM allows for the modeling of spatial interactions between the built, transportation, and utility infrastructures, thereby creating opportunities for evidence-based spatial decisions. For example, simulations based on CIM can assess the implications of new road alignments or housing developments for accessibility, energy use, or flood risk. In terms of infrastructure management, CIM allows for the data-enabled monitoring, maintenance, and upgrading of the existing asset base. Finally, CIM assists in environmental assessment by relating urban morphology to resource flows, including water, waste, and emissions.

A noteworthy finding from the review of Souza and Bueno is the acknowledgement of CIM as a strategic tool for sustainable development. The authors state that cities can integrate CIM with Geographic Information Systems (GIS), as well as their environmental databases, to measure their ecological footprint, plan low-carbon mobility systems, and systematically monitor resilience indicators. In this sense, CIM assists directly in the implementation of the United Nations Sustainable Development Goals (SDGs), specifically SDG 9 (Industry, Innovation and Infrastructure) and SDG 11 (Sustainable Cities and Communities).

Nonetheless, while the possibilities for CIM are promising, Souza and Bueno (2022) also highlight challenges to CIM implementation. The most frequent problems are associated with data interoperability, institutional coordination and bureaucracy, and human resource capacity. Many municipalities still work with old or fragmented datasets, which makes it difficult to integrate BIM, GIS and IoT. Additionally, while data formats could play an important role, the lack of harmonized data formats between urban sectors limits collaborative work. Furthermore, they argue operationalizing the use of open standards such as CityGML, LandInfra, and IFC 4x will improve both the compatibility of data between urban sectors and the sustainability of that data.

The review points out organizational barriers as well. CIM projects often require cross-sectoral collaboration among public administrations, private consultants, and software providers, with institutional silos being a continuing barrier. Moving from the traditional 2D environment to integrated 3D/4D/5D developed environments necessitates reporting and investment, not just technology, but also transforms in organization and training. In developing countries, they face insurmountable barriers including limited financing and lack of digital connectivity/infrastructure. Hence, CIM is primarily deployed in high-income countries with its uses and potential in African and Latin American countries remaining largely unexplored.

Another key takeaway from Souza and Bueno's analysis is the demand for data to be consistently updated and maintained as an important part of ensuring CIM models are accurate and pertinent. CIM does not operate like static data GIS systems; CIM relies on a real-time dynamic information flow that links back to changes in the urban environment. New technologies, such as LiDAR scanning, UAV photogrammetry, and the design of sensor networks, will be integral to this process. These novel technologies enable automatic data tapping and integration of terrain, traffic, and environmental data into the base CIM model, transforming the CIM model into a living digital twin of the city.

Moreover, the authors stress the need for collaborative platforms to properly execute CIM. Online cloud-based environments—like Autodesk Construction Cloud or Bentley OpenCities—allow stakeholders to access, see, and edit a model, at the same time. This improves transparency, coordination, and accountability during the project life cycle. According to Souza and Bueno (2022), these platforms are highly critical for urban projects at a large scale, where disciplines must operate concurrently on shared data.

Moreover, the review addresses the merger of CIM with the new digital technologies such as Artificial Intelligence (AI), machine learning, and blockchain. AI methods can facilitate object recognition and data categorization in CIM models, thus increasing efficiency and reliability. In comparison, blockchain can also provide data integrity and traceability in joint projects by logging each modification in an unchangeable ledger. The combination of these developments indicated the next stage in the evolution of CIM toward fully autonomous and self-updating urban systems.

In terms of methodology, Souza and Bueno's review is systematic in nature and analyzes 150 scholarly papers published between 2010 and 2021. They also provide a bibliometric review of countries in the same time frame, noting there has been a substantive increase in papers relating to CIM research from 2016, which correlates with the increase in interest in the digital transformation of cities. Most of the studies are emerging from Europe and Asia (particularly China, South Korea, and the UK) who have national or country-wide standards for digital cities. Conversely, Africa and South America are notably absent highlighting the need for situated case studies which could leverage the CIM frameworks but be situated within their context and limitations on resources. Ultimately, the review finds that CIM is an effective means for pursuing data-driven urban governance. CIM combines geospatial modeling, infrastructure simulation, and collaborative management and facilitates the shift from reactive urban management to proactive planning. Realizing this transformation will demand a long-term commitment by policy makers, interdisciplinary collaboration, and capacity building among municipal and academic representatives. The authors further suggest, for sustainable and equitable digitalization of cities, that principles of CIM must be incorporated into urban master plans and planning for infrastructure and that training for professionals must be renewed.

CONCLUSION

Overall, this investigation looked at the possibility of using City Information Modeling (CIM) to improve road extra to the housing under construction for SIC Access in Ekoko II of Mfou commune. Specifically, the studies goal was to obtain a detailed 3D CIM model of the access road, with all geometric and functional elements presented, to serve as a reference for subsequent phases of the project. The findings indicate that a CIM based project designed and delivered a significant amount of added value (collaboration) throughout the road project lifecycle. The creation of a detailed 3D model in conjunction with the intended 4D (planning) and 5D (cost estimation) functions provided a greater amount of knowledge as well as a coherent overview of the project. This also provided greater coordination throughout the project with stakeholders. The developed model improved the quality of design, and significantly reduced the risk of the project exceeding anticipated costs and schedules. The platform developed for the project included a CIM based collaboration that provided several means of communication and information that was necessary for the various stakeholders and participants. Monitoring the progress of the project in real time was also possible, since communication was instant, and potential problems could be identified rapidly. The outcomes of this research illustrate the significance of bringing CIM into the design and construction of road infrastructure. There are many ways that this technology can enhance the efficiency, quality, and sustainability of all projects, but without proper training for the construction professionals and clear collaborative solutions, if successful, its application will not achieve widespread adoption. The future is bright. The continued increase in technology has the potential for enriching CIM and its usefulness on the site, especially in simulation or data analysis. We believe it would be helpful to compare to the existing costs of construction to our CIM offer. The benefit of CIM technology is partnership with an emerging technology, virtual and augmented reality, as well as the Internet of Things, which will provide a new lens through which to visualize and engage with the process or in interactive experience models.

REFERENCES

Ait-Lamallam, S., Sebari, I., Yaagoubi, R., & Doukari, O. (2019). Vers une extension du Building Information Modeling aux projets d'infrastructure routière: Cas d'étude d'un échangeur autoroutier au Maroc. *Revue Marocaine des Sciences Agronomiques et Vétérinaires*, *28*(1), 609–617.

Akob, Z., Hipni, M. Z. A., & Abd Razak, A. A. A. (2019). Deployment of GIS + BIM in the construction of Pan Borneo Highway Sarawak, Malaysia. [IOP Publishing.]. *IOP Conference Series. Materials Science and Engineering*, *512*, 012037. DOI: 10.1088/1757-899X/512/1/012037

Alizadeh, K., Bajgiran, M., & Nik, E. (2016). Building Information Modeling (BIM): A study to prioritize applications, risks and challenges. *International Conference on Industrial Engineering and Operations Management*, (18), 1–7.

Beil, C., & Kolbe, T. H. (2017). CityGML and the streets of New York: A proposal for detailed street space modelling. In *Proceedings of the 12th International 3D GeoInfo Conference 2017*, 9–16. DOI: 10.5194/isprs-annals-IV-4-W5-9-2017

Berlato, M., Binni, L., Durmus, D., Gatto, C., Giusti, L., Massari, A., Toldo, B. M., Cascone, S., & Mirarchi, C. (2025). Digital platforms for the built environment: A systematic review across sectors and scales. *Buildings (Basel, Switzerland)*, *15*(14), 2432. DOI: 10.3390/buildings15142432

Castañeda, K., Sánchez, O., Herrera, R. F., Gómez-Cabrera, A., & Mejía, G. (2024). Building Information Modeling uses and complementary technologies in road projects: A systematic review. *Buildings (Basel, Switzerland)*, *14*(3), 563. DOI: 10.3390/buildings14030563

Dantas, H., Sousa, J., & Melo, C. (2019). The importance of city information modeling (CIM) for cities' sustainability. [IOP Publishing.]. *IOP Conference Series. Earth and Environmental Science*, *225*, 012074. DOI: 10.1088/1755-1315/225/1/012074

Das, D. K. (2024). Exploring the symbiotic relationship between digital transformation, infrastructure, service delivery, and governance for smart sustainable cities. *Smart Cities*, *7*(2), 806–835. DOI: 10.3390/smartcities7020034

Deng, Y., Cheng, J. C. P., & Anumba, C. (2016). A framework for 3D traffic noise mapping using data from BIM and GIS integration. *Structure and Infrastructure Engineering*, *12*(10), 1267–1280. DOI: 10.1080/15732479.2015.1110603

Deprêtre, A., Mielniczek, A., & Jacquinod, F. (2023). Le City Information Modelling (CIM) au service d'un projet urbain: Retour d'expérience sur la première phase de mise en œuvre du CIM d'un quartier. *Flux*, *3*(133), 57–75. DOI: 10.3917/flux1.133.0057

Gikas, V. (2012). Three-dimensional laser scanning for geometry documentation and construction management of highway tunnels during excavation. *Sensors (Basel)*, *12*(8), 11249–11270. DOI: 10.3390/s120811249 PMID: 23112655

Gil, J., Almeida, J., & Duarte, J. P. (2011). The backbone of a city information model (CIM). In *Respecting Fragile Places: Education in Computer Aided Architectural Design in Europe*, 143–151.

Irschick, D. J., Christiansen, F., Hammerschlag, N., Martin, J., Madsen, P. T., Wyneken, J., Brooks, A., Gleiss, A., Fossette, S., Siler, C., Gamble, T., Fish, F., Siebert, U., Patel, J., Xu, Z., Kalogerakis, E., Medina, J., Mukherji, A., Mandica, M., & Lauder, G. (2022). 3D visualization processes for recreating and studying organismal form. *iScience*, *25*(9), 104867. DOI: 10.1016/j.isci.2022.104867 PMID: 36060053

Jovanović, D., Milovanov, S., Ruskovski, I., Govedarica, M., Sladić, D., Radulović, A., & Pajić, V. (2020). Building virtual 3D city model for smart cities applications: A case study on campus area of the University of Novi Sad. *ISPRS International Journal of Geo-Information*, *9*(8), 476. DOI: 10.3390/ijgi9080476

Khoshamadi, N., Banihashemi, S., Poshdar, M., Abbasianjahromi, H., Tabadkani, A., & Hajirasouli, A. (2023). Parametric and generative mechanisms for infrastructure projects. *Automation in Construction*, *154*, 104968. DOI: 10.1016/j.autcon.2023.104968

Lasorella, M., & Cantatore, E. (2025). CityGML-based model for the recovery and management of built cultural heritage: A systematic review towards digitalized processes. City. *Territory and Architecture*, *12*(1), 10. DOI: 10.1186/s40410-025-00259-7

Najafi, P., Mohammadi, M., van Wesemael, P., & Le Blanc, P. M. (2023). A user-centred virtual city information model for inclusive community design: State of art. *Cities (London, England)*, *134*, 104203. DOI: 10.1016/j.cities.2023.104203

Nchang, L. N., Fombe, L. F., & Wanie, C. M. (2024). Population vulnerability to the state of public transport in Bamenda city, Cameroon. *ISAR Journal of Arts. Humanities and Social Sciences*, *2*(10), 85–98.

Nijkamp, P., Kourtit, K., Krugman, P., & Moreno, C. (2024). Old wisdom and the new economic geography: Managing uncertainty in 21st century regional and urban development. *Regional Science Policy & Practice*, *16*(10), 100124. DOI: 10.1016/j.rspp.2024.100124

Norrman, K. E. (2023). World population growth: A once and future global concern. *WORLD (Oakland, Calif.)*, *4*(4), 684–697. DOI: 10.3390/world4040043

Qi, Y., Liu, C., Scaioni, M., Li, Y., Qiao, Y., Ma, X., Wu, H., Zhang, K., & Wang, D. (2024). Geometric information constraint 3D object detection from LiDAR point cloud for autonomous vehicles under adverse weather. *Transportation Research Part C, Emerging Technologies*, *161*, 104555. DOI: 10.1016/j.trc.2024.104555

Rezvani, S. M. H. S., Falcão, M. J., Komljenovic, D., & de Almeida, N. M. (2023). A systematic literature review on urban resilience enabled with asset and disaster risk management approaches and GIS-based decision support tools. *Applied Sciences (Basel, Switzerland)*, *13*(4), 2223. DOI: 10.3390/app13042223

Shamaee, S. H., Yousefi, H., & Zahedi, R. (2024). Assessing urban development indicators for environmental sustainability. *Discover Sustainability*, *5*(1), 341. DOI: 10.1007/s43621-024-00563-1

Sharma, N., Kumar, R., Vihari, N. S., Arora, M., & Saini, J. R. (2025). Prioritizing smart city themes for multi-national enterprises and United Nations Sustainable Development Goals. *Sustainability (Basel)*, *17*(10), 4251. DOI: 10.3390/su17104251

Šimenić, D. (2021). Building Information Modeling (BIM) for road infrastructure: TEM requirements and recommendations. Genève (Suisse), 119 p.

Souza, L., & Bueno, C. (2022). City information modelling as a support decision tool for planning and management of cities: A systematic literature review and bibliometric analysis. *Building and Environment*, *207*, 108403. DOI: 10.1016/j.buildenv.2021.108403

Taheri, A., & Sobanjo, J. (2024). Civil Integrated Management (CIM) for advanced level applications to transportation infrastructure: A state-of-the-art review. *Infrastructures*, *9*(6), 90. DOI: 10.3390/infrastructures9060090

Takva, Ç., Top, S. M., Gökgöz, B. İ., Gebel, Ş., İlerisoy, Z. Y., İlcan, H., & Şahmaran, M. (2024). Applicability of 3D concrete printing technology in building construction with different architectural design decisions in housing. *Journal of Building Engineering*, *98*, 111257. DOI: 10.1016/j.jobe.2024.111257

Wang, H., Pan, Y., & Luo, X. (2019). Integration of BIM and GIS in sustainable built environment: A review and bibliometric analysis. *Automation in Construction, 103*, 41–52. DOI: 10.1016/j.autcon.2019.03.005

Xu, X., Ding, L., Luo, H., & Ma, L. (2014). From building information modeling to city information modeling. [ITcon]. *Journal of Information Technology in Construction, 19*, 292–307.

Xu, Z., Qi, M., Wu, Y., Hao, X., & Yang, Y. (2021). City information modeling: State of the art. *Applied Sciences (Basel, Switzerland), 11*(19), 9333. DOI: 10.3390/app11199333

Yu, W., Zhou, X., Wang, D., & Dong, J. (2025). The development and construction of City Information Modeling (CIM): A survey from data perspective. *Applied Sciences (Basel, Switzerland), 15*(9), 4696. DOI: 10.3390/app15094696

Chapter 5
Leveraging GIS for Intelligent Traffic Flow and Public Transport Planning:
GeoSmart Mobility

Sitesh Kumar Singh
https://orcid.org/0000-0002-7108-0808
College of Engineering, National University of Science and Technology, Oman

Jaishri Gothania
https://orcid.org/0000-0002-2656-642X
Lingaya's Vidyapeeth, India

Leevesh Kumar
https://orcid.org/0000-0001-9476-0841
World University of Bangladesh, Bangladesh

ABSTRACT

In this chapter, the paradigm shifting Geographic Information Systems (GIS) in facilitating the intelligent urban mobility is discussed. Since its development as a mapping instrument, GIS has undergone numerous advancements on the one hand and challenges the other to become a part of traffic, multimodal networks, and public transport planning and management in cities. The chapter focuses on GIS use concerning congestion studies, optimization of routes and places a stress on spatial equity, citing global examples of GIS use in Singapore, Amsterdam, India and the U.S. It addresses the questions like the data privacy, interoperability, institutional

DOI: 10.4018/979-8-3373-4058-6.ch005

barriers and leverages open data, Digital twins, and strategic roadmaps. The conclusion of the discussion frames GIS as one of the building blocks of future-proof, resilient, and inclusive urban transportation systems.

INTRODUCTION TO GIS IN URBAN TRANSPORT

Evolution of GIS: From Cartography to Smart City Tool

Geographic Information Systems (GIS) have since evolved over the past few decades to become more complex than just the basic structure of a cartographic analysis and into an effective analytical platform which forms the backbone of the city development and transport sector. The initial applications of GIS were restricted initially as they merely superimposed data on the two-dimensional plane even though it was used to map and draw the static spatial representation. However, with the advent of computational advances and introduction of numerical data to the Internet and real-time, GIS was converted into a multidimensional format, which was also able to provide real-time and interactive forecasting and spatial analysis of the data (Xhafa & Kosovrasti, 2015). This trend is what has emerged together with the introduction of the idea of smarter cities where various digital infrastructures, including transport systems rely on spatial intelligence to execute their services effectively.

Figure 1. Evolution of GIS in Urban Transport

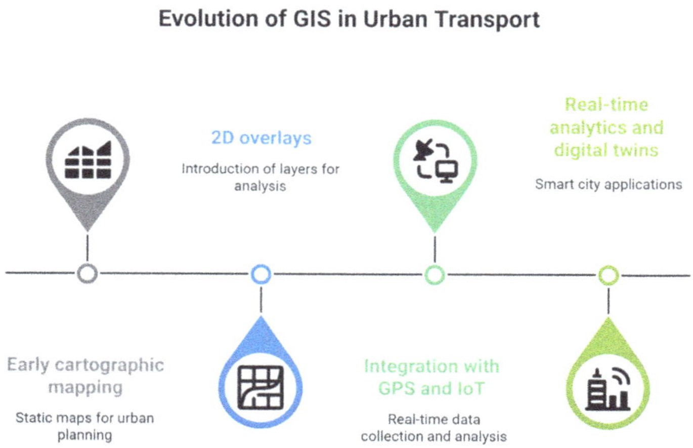

The modern GIS systems integrate the data that are obtained in various ways, such as GPS sensors, IoT-interconnected infrastructure, satellite imagery, mobile sensors, to provide systems of urban mobility with more active and detailed information (Zaroujtaghi et al., 2025). These capabilities are not exclusive to mapping anymore; they can be used to support a decision-making using data modelling, predictive analytics and simulation, and, therefore, create a connection between real infrastructure and the digital governance. In addition to technological development, institutional capacity in the governance of urban transport has been re-established through the development of GIS into frameworks of digital twins. Connecting real time spatial data with predictive and analytic dashboards allow authorities to visualize the result of policy before implementation to enhance transparency and accountability. With this transformation, policymaking is now informed by data and continuous spatial feedback instead of the conventional reports, which enhances more adaptability and evidence-based planning. Therefore, GIS is an essential enabler of digital twins of urban transport systems, much more adaptive, flexible, and people-oriented regarding the planning processes. Figure 1 shows the evolution of GIS capabilities through conventional mapping instruments to dynamic digital twins that are utilized in the modern urban mobility platform.

GIS is used in smart city applications as a communication point joining the information of the transport department, urban planning department, environmental monitoring department, and emergency service department. This will facilitate optimization of the entire transportation system in highly populated and rapidly expanding towns. It also can help in coordination among different stakeholders as it visualizes the common spatial information in readable format and then makes decisions (Aljoufie, 2013). With the emergence of bigger-picture technologies like GIS, the adoption of such technologies has exploded, and the cause is that the GIS can serve as a strategic tool to drive integrated city mobility. Although developed nations are using the developed spatial data infrastructures and institutional frameworks to actualize GIS-based governance, most developing countries continue to grapple with their disjointed datasets, weak technical capacity, and uneven geospatial policies. This results in the process of change in moving cartographic mapping to spatial intelligence being uneven due to differences in digital preparedness and combining policies.

Importance of Spatial Intelligence in Transport Planning

The good thing about the use of spatial intelligence in transport planning is that it provides the interaction between the infrastructure and the population dynamics and the travel behaviour as well as the environmental conditions. Transportation is spatial; the plan should possess precise conception of the level of congestion, human flow, the area that is over-supplied as well as undersupplied. GIS enables the planners to view these patterns in space and time which is a dynamic filter through which they make operating and long-term decisions regarding them (Xhafa & Kosovrasti, 2015).

The spatial intelligence adds to the fertile functionality which are the route optimization, accessibility analysis, infrastructure prioritization and the multimodal integration. The GIS-based equity-based planning can also be focused on the disadvantaged population or geographically remote population, which will further make the transport services more inclusive and address the needs of the target population by including the spatial data to demographic and socioeconomic data (Zhou et al., 2021). The GIS is also used by the planners to find out the environmental impact of transport decisions through overlaying transport data on transport emissions on the land use and health indicators as a way of ensuring transport strategies are aligned with the sustainability orientations.

India and other developing economies are also using GIS to address real time urban problems. Singh et al. (n. d.) presented as an example that the models of machine learning in combination with GIS can forecast air quality along the overcrowded urban streets and therefore it can act as feedback between transport infrastructure and environmental health. Similarly, automated means that use GIS within the urban traffic control centres may be used to monitor the current congestion systems in real-time, operate and respond to incidences, and conduct dynamic signal control that enhance efficiency and safety about mobility (Shi et al., 2008).

The complexity of urban settlements is on the rise, so the role of a spatial intelligence cannot be disregarded in the data-driven and sustainable mobility planning. It converts the transport systems that are reactive and manually operated in nature to the proactive and scenario-based mechanisms. In this way, GIS ceases being a secondary technology; it is a major component of the contemporary urban governing system, which offers the mechanisms of creating resilient, inclusive and future-proof transportation systems. These transformative results are emphasized by the empirical evidence. GIS-based real-time bus management systems in Seoul minimized the waiting time of passengers by approximately 15% and enhanced the network efficiency (Bachechi et al., 2021). Likewise, Helsinki incorporates the GIS-based mobility dashboards in its digital twin platform to organize multimodal activities and streamline the route planning (Liu et al., 2023). In Curitiba, land-use and transport integration developed through GIS enhanced access mapping, which led to fair bus rapid transit expansions (Papa & Bertolini, 2015). These instances illustrate that GIS is not simply an abstract enabling tool but it is an established process of operational effectiveness and spatially conscious governance in a variety of urban situations.

The existence of regional inequalities is a major factor in influencing the use of GIS in promoting an equitable transport planning process. Spatial datasets have been developed by high-resolution, and advanced demographic modelling allows planners in developed economies to measure accessibility disparities precisely and conduct inclusive policy interventions. Conversely, emerging economies are normally characterized by issues that include a lack of complete data, informal modes of transport and poor institutional coordination, which limits the accuracy of the model. However, creative applications of open-source GIS and participative mapping in African, Asian, and Latin American urban centres are showing increased possibilities of localizing equity-based fixations of mobility and modifying international approaches to regional socioeconomic circumstance.

GIS applications in climate resilience and sustainable mobility

Since the urban transport systems have been experiencing growing climate-related upheavals, mobility planning has been compelled to incorporate resilience thinking. GIS is the tool of analytical connection between spatial intelligence and resilience governance, as it helps to locate weak points, model risk situations, and prioritize adaptive infrastructure investments. The Urban Climate Resilience as laid out in the Urban Climate Resilience Guidelines (2023) by the UN-Habitat emphasize the potential of geospatial data in the prediction of flood, heat, and emission trends influencing the transport functions. Urban adaptation studies developed by IPCC Working Group II (2022) provide insights into this matter. Therefore, GIS-based governance turns climate resilience into an emergency recovery initiative to an opportunity-driven part of the urban mobility optimization and long-term sustainability.

Geographic Information Systems (GIS) are essential in incorporating climate resiliency in transport planning in cities and promoting sustainable mobility. When topography is overlaid by spatial data of land use, infrastructures, and environmental risks, GIS can be used by planners to evaluate the danger of flooding roads or heat islands or any other place that is vulnerable to extreme weather conditions. This leads in advance planning to enhance a robust transport system that stands the recovery and ensures a small fraction of downtime of services in case of climate incidents. As an example, flood modelling based on the use of GIS can be used to find alternative evacuation paths and focus on the strengthening of infrastructure in the most vulnerable areas. Simultaneously, GIS enhances sustainable movement through the optimization of multimodal transport systems. Geospatial analysis of real-time data can draw attention to the underserved areas, allowing us to provide an equal distribution of bus stops, cycling lanes, and pedestrian routes. Combining traffic flow information with air quality and emissions maps also enables policymakers to estimate the environmental effect of transport policies so that the mobility solutions minimize the emissions of greenhouse gases. Also, GIS can be used to model a scenario, low-carbon futures, in which cities can model the value of electric buses, smart parking, or car-free zones. Finally, GIS is a decision-support tool that can be used to meet the requirements of climate adaptation and sustainable transport goals to establish a resilient and environmentally responsible urban mobility system.

GIS FOR TRAFFIC FLOW ANALYSIS

Congestion Monitoring and Hotspot Detection

The problem of urban congestion is one of the most apparent and ancient problems of transport planning. Traffic in urban areas poses a challenge of observing traffic changes and complexity through the application of traditional monitoring devices which cannot capture the dynamics and dynamics of the newly emerging mobility patterns. GIS has the ability of displaying advanced spatial systems to identify, map and analyse points of unruly traffic, and integrate numerous streams of data both in space and time.

GIS can also enable the urban planner to undertake mapping of the congestion patterns by overlaying the real time and past-based information of speed on a regular basis and other travel time and the capacity of the roads as well. The gathering of these spatial data can be in the form of loop detectors, Bluetooth level sensors, phones and automatic vehicle location. The geographical mapping of traffic density aids in determining the bottlenecks, the busy-hours of the networks and the delays, which recur throughout the networks (Solé-Ribalta et al., 2016). To give an example, it could be possible to locate the arterial roads with constant delays with the help of GIS-based heatmaps and manipulate infrastructure adjustments and drives on roads.

Also, it is possible to analyse these phenomena using GIS and apply spatial clustering methods such as kernel density estimation (KDE) and hot spot analysis to statistically determine where there is high intensity of congestion. These tools are not only capable of providing descriptive data but also allow taking active action when it comes to policymaking in terms of quantifying the level of congestion and correlating it with land-use, socio-economic and environmental factors (Solé-Ribalta et al., 2016). This space intelligence can be applied in the allocation of limited resources in a rational manner such that the intervention can be data and context based.

Real-Time Data from IoT and GPS

The real power of GIS in traffic analysis is its ability to use real-time sources of data which include Internet of Things (IoT) sensors and Global Positioning System (GPS) devices. Intelligent infrastructure- Traffic lights, vehicle sensors and CCTV analytics, feed the data on continuous location and movement into GIS dashboards. These dynamic data sets enable the transport authorities to identify abnormalities like spikes in congestion, accidents, and illegal use of a lane or to use fast measures by controlling signals dynamically or optimizing routes. Real-time geospatial monitoring therefore changes traffic management to be based on observing rather than responsive.

GPS of taxis, buses on the highway, delivery and other vehicles, even individuals when they are on their way to work every day can update the information about the place continuously which can be included in the data layers of the GIS. These live geolocation data may be welcomed to compute valuable mobility routes such as vehicle flows, change in speed and stop times, which are essential in the knowledge of the micro level traffic behaviour (Liu et al., 2023). GIS dashboards, connected to GPS, are implemented in such cities as Seoul and Singapore to monitor their congestion on a near real-time basis, enabling redirection of buses and emergency vehicles, and hence, optimizing their operation and the commuter experience. The architecture of a GIS-based traffic monitoring system as shown in Figure 2 presents the idea of using real-time IoT and GPS data to analyse and present them in analytical dashboards.

Figure 2. GIS-Enabled Traffic Flow Monitoring Architecture

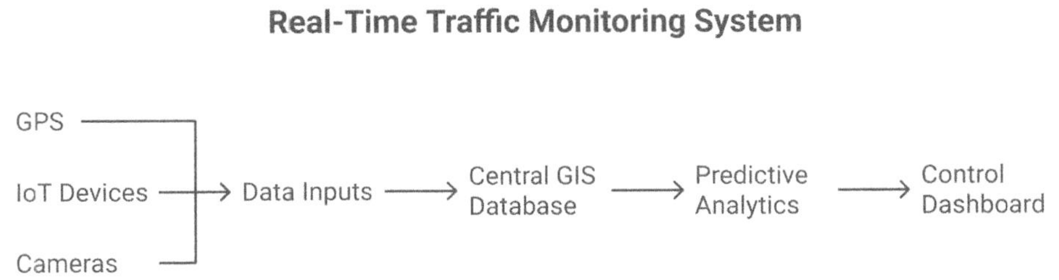

Markedly, real-time data with GIS also envisages dynamic mapping interface that can be delivered via the web-based or mobile platform. This will be made possible through allowing the city planners, emergency responders, and even the citizens, to receive and respond to real-time traffic data, which will be used to develop a more reactive and transparent system of movement control (Bachechi et al., 2021).

Predictive Modelling, Dashboards, Simulation Tools

Using real-time integration, predictive GIS analytics can transform spatial data into future insights that inform the transport policies. Using a mixture of historical traffic data, socio-economic information, and environmental factors, it is possible to identify the effects of congestion pricing, test the viability of new transit routes, or predict the potential emission cuts in alternative mobility conditions. The SUMO, AI-enhanced GIS tools enable authorities to simulate the consequences of the regulatory decisions prior to their execution. As illustrated by cities like London and Singapore, foresight on urban mobility is implemented using predictive dashboards, which can change and adapt the rate of tolling or the priority of traffic signals in real time, which prove to be more equitable and efficient governance of urban mobility.

The Simulation of Urban Mobility (SUMO) is a traffic simulator provided as open-source and capable of using GIS platforms to conduct the spatial modelling of one of the most powerful tools in this domain. SUMO simulates the impact of many changes on traffic conditions using the input of road geometries, traffic signalling timing, vehicle flow, that influences how traffic behaviours of the road (Alkaissi, 2024). This type of simulation will enable the planners to have a visual picture of what will occur once several interventions have been put in place to perfect the transport interventions.

A lot of the problems mentioned above can be applied to the situation in India, and Singh and Saraswat (2019) clarified that simulations based on GIS can determine the results of mid-sized cities in India to comprehend and control any traffic pattern more effectively using the SUMO and real traffic flows. The analysis carried out by them revealed that it is important to integrate survey data on the ground and GIS information as input to make the resultant model have the contextual accuracy and therefore more useful in grids of traffic programs within a city.

GIS FOR PUBLIC TRANSPORT PLANNING

Network Design and Route Optimization

An accessible and sustainable urban mobility is fundamental in the establishment of efficient networks of public transport. Geographic Information Systems (GIS) plays a key role in this process since they enable the planners to visualize, analyse and optimize transport routes using the spherically and demographically relevant data and the data useful to the operation. With an intervention like GIS integrated into the transport network modelling, cities can produce data-based layouts that make the least use of redundancy and best use of coverage and travel times.

GIS can deliver spatial images of the transit routes and data layers that are fed in such as population density, land use, work clusters, travel demand behaviours. These insights may assist in deciding where to expand or merge the routes by the planners to enhance their knowledge concerning the gaps and overlaps of the services and make improved decisions (Schiller & Kenworthy, 2017). GIS can also be used in the optimization of routes using shortest path algorithms, networks topology and transfer minimization models. The performance of the routes is measured by spatial, which evaluates the time taken to travel, area covered, indices of accessibility etc., must have the capability of ensuring the planners the optimal arrangements that perform effectively and at the lowest possible costs.

The GIS has also been crucial in rationalization of the public bus service of cities like Pune, Ahmedabad in India so that it is made sure that the network runs better and the cost of operations is reduced. The review of Singh and Bansal (2021) of the SME data on urban growth trends and traffic jams using the GIS illustrated how to reorganize the current paths of the urban transport system to enhance multimodal connectivity and optimization of the fleet results.

Accessibility Analysis and Equity Considerations

Besides technical efficiency, GIS will provide the necessary tools in the decision to achieve spatial equity of the transport systems. The accessibility analysis of GIS helps in establishing the areas where access to essential facilities such as hospitals, schools, and workplaces is low. The information supplied with social-economic information, e.g., of income/gender or age distribution, can be overlaid with the route of transport, in such a way that the underserved can comprise the marginalized/ disadvantaged populations, availed by planners. As an example, an index of accessibility based on GIS distinguishing between central and peripheral areas in Pune found that the average travel time to a workplace was 35 percent higher in central areas than in peripheral areas, and spatial equity mapping in Brazil found under-serviced districts having less than 60 percent of their areas covered by public transport. These quantitative examples improve the equity evaluation and evidence-based mobility planning.

Cumulative opportunity, manipulation by forces of gravity, isochrone mapping are measurements of accessibility that have been extensively applied in GIS to calculate the ease by which a person can reach places of desired destinations by using a public transport (Delbosc & Currie, 2011). Those tools serve to identify Transit deserts, the areas that lack reasonable access to any transport system available and help in directing efforts that are expected to help restrain spatial exclusion.

Moreover, the longitudinal equity audits are feasible to be achieved since the GIS will be able to monitor the effects of the change of infrastructure on the availability over the long term. This aspect is critical in fast evolving cities in terms of its demographics or land-use. To illustrate, in the Brazilian city of Sao Paulo, GIS-based equity estimates revealed that a systematic under-investment had been made in the low-income peripheries in the city in options such as public transport and that the calculus led the city to redistribute funds and prioritize the need to include planning (Du, 2024).

Figure 3. Accessibility Mapping for Equitable Transport Planning

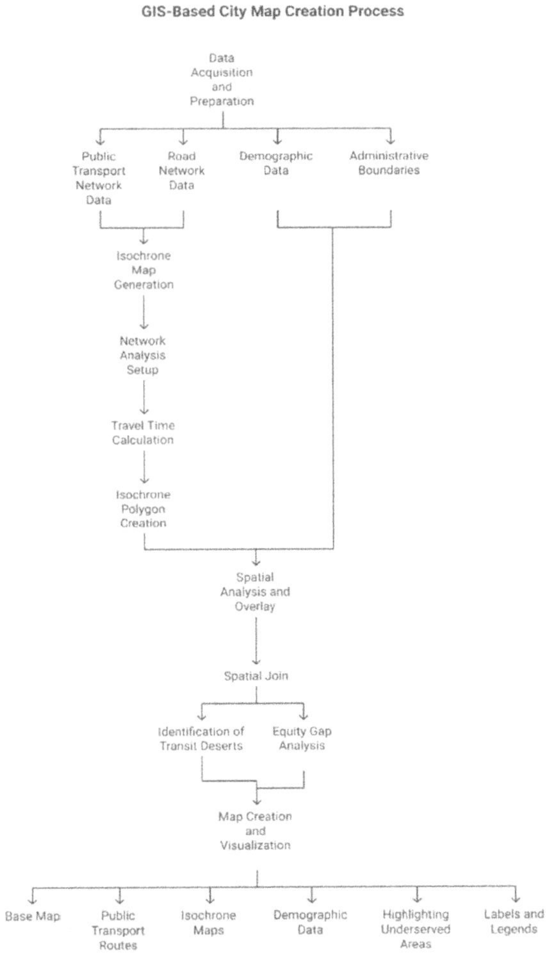

Singh et al. (2025) transfer it to the same lenses by applying GIS-incorporated models to assess, regarding the effect of traffic congestion in low-income neighbourhoods, both environmental and human health. What their findings showed is that spatial intelligence is inestimable in uncovering different unbalanced weights of bad transport planning on susceptible groups. An example of accessibility mapping GIS based is provided in figure 3, where spatial inequities in coverage of the public service transport can be identified.

Integration of Modes (Bus, Metro, Cycling, Walkability)

The ability to offer multimodal integration is one of the functions of GIS in the planning of transport as it currently is. The seamless movement of different modes of transportation, i.e., buses, metros, cycling tracks and pedestrian tracks is significant in providing a highly organized mobility system, which is efficient, user-friendly and sustainable. GIS assists the planners to visualize and analyse multimodal networks, the territories where spatial gaps, integrations, and destinations of transfers occur, which has an impact on user experience and system performance.

GIS is also capable of identifying the locations of such physical or time-based disconnections among modes by mapping such infrastructure as bus stops, metro stations, bike-sharing docks, the location of walking or bicycle lanes, etc. (Papa & Bertolini, 2015). These insights will be part of the planning of such strategies as the first mile/last-mile connectivity improvements, planning park-and-ride, construction of mobility hubs, which will serve as the points of smooth transition.

The other useful tool in multimodal integration is the analysis of walking, which is performed using GIS. Spatial indicators such as land-use mixture, the directness of walking paths, and the concentration of crossings may help the planners to evaluate the accessibility and safety of walking conditions in the areas that surround the transit nodes. The GIS-based programs will be helpful in pinpointing the vulnerabilities of the infrastructures i.e. the lack of sidewalks, unsafe crossing or low lighting among many others- which are the typical elements that deter active movement.

Other instances of the cities that have utilized GIS in integrated cycling-transit planning include Amsterdam where bike roads are to be laid depending on the historical data of the cyclists movement, mode of using the public transport utilization and the land-use pattern amongst others where the cycling route is determined by the movement of the cyclists in addition to the utilization of the public transport. The analyses apply to the dynamic bike-lane assignment, prediction and prioritization of the infrastructure in modal shift. GIS can also support the intermodal transfer time and cross-modal connection time viability assessment (Elgayar and Jrade, 2022).

The objective behind the implementation of integrated mobility platforms that bundle metro, bus, and non-motorized transport information on GIS platforms has slowly become the fad in the Indian cities. The convergence facilitates such aspects as a single ticket, cross-modal travel planning and the development of inclusive transport systems which help push towards a less reliance on their personal vehicle.

Table 1. Comparative Overview of GIS Applications Across Urban Mobility Domains

Urban Mobility Function	Key GIS Application	Data Sources Involved	Tools/Techniques Used	Expected Outcomes
Traffic Flow Management	Congestion monitoring, hotspot detection	IoT, GPS, CCTV, loop detectors	Heatmaps, KDE, real-time mapping	Bottleneck identification, dynamic control
Public Transport Planning	Route optimization, network design	Ridership data, land use, census data	Network analysis, accessibility mapping	Efficient coverage, reduced redundancy
Accessibility & Equity	Transit deserts identification	Socioeconomic, demographic, transit	Isochrone mapping, equity audits	Inclusive planning, service equity
Multimodal Integration	First/last mile connectivity, transfer points	Infrastructure layers, travel data	Node mapping, walkability indices	Seamless commuting, mode shift

Table 1 below includes a comparative summary of GIS-enabled applications through major urban mobility areas, which demonstrates the variability of data, tools, and anticipated results that have been used to lead to smart transport planning. Although these exist, multimodal coordination in the developing context has had institutional fragmentation, incompatible data systems and behavioral inertia of commuters and operators. The absence of a strong inter-agency governance, the lack of real-time data sharing, and resistance of users to a modality change frequently delay the seamless integration, which is why it is necessary to introduce unified frameworks and transport governance reforms.

INTEGRATION WITH SMART TECHNOLOGIES

GIS and ITS Synergy

The Intelligent Transport Systems (ITS) and Geographic Information Systems (GIS) combination is among the significant advances into the construction of smart urban mobility. Where ITS technologies focus to collect, process, and distribute transportation-based data using sensors, controllers, and communication platforms, GIS will deliver the data with spatial attributes and abilities of analysis. All of them create a synergistic platform that can adequately handle traffic both in real time and plan and optimize infrastructure and the system.

GIS provides ITS structures with the ability to transcend the condensed perception of data by assigning it places, signal conditions, environmental conditions all into one geospatial framework. This integration makes it possible to use it in such tasks as the adaptive control of traffic lights, the assigning of the route dynamically, the identification of incidents, and the establishment of priorities to vehicles in GE (Alsubai et al., 2024). As an illustration, signs that have GIS-based ITS dashboards embedded may display real time maps of the traffic densities, signal phases and stream of traffic that enable traffic or incident control centers (TMCs) to respond dynamically to road situations.

Such cities as London and Singapore have used GIS and ITS to enable a centralized management of the urban mobility system, where real-time sensor information is mapped and interpreted to overcome congestion, plan transport and demand-responsive transport (Liang et al., 2021). In the Indian case, GIS-based ITS has been tried by Delhi Transport Corporation (DTC), to optimize bus fleet operation based on real-time ridership, and congestion information.

Crowdsourcing, Mobile Apps, AI-Enhanced Analysis

With the introduction of smartphones, wearable apps, and location-based applications, a new horizon has been explored on how crowdsourced data can be included in GIS applications. GIS plays a significant role in production of other values whereby transmission of public contribution through mobile applications which includes the timing of travel, congestion warnings, accident reports and user opinions enhances the existing GIS information with high frequency user centric information. These inputs will be suitable in making improved real-time traffic maps, service quality analytics, and behavior data.

Crowdsourcing has particularly been of value in the cities where sensors are not well installed. The user contributed information is availed to applications like Waze and Google Maps who thus offer near real time information regarding the state of the roads through congestion, roadblock, or delays. The data when augmented in GIS systems enhances situational awareness on both sides of the planners and the operators (Huang et al., 2024). These technologies together are complements of GIS functions, as opposed to being duplicates. ITS and IoT networks produce a continuous stream of data, whereas GIS enables the spatial intelligence framework that places the data in perspective of what can be done to plan. The crowdsourced contributions get added to the geographic areas to increase the user point of view and the AI models enhance the accuracy of predictions, and both work together to build a synergistic ecosystem, with GIS as the master system of perception, analysis, and governance.

The further extension of the capability of Artificial Intelligence (AI) in the analysis is due to its integration with GIS. Machine learning algorithms may be employed with the geospatial dataset to reveal the buried trends in commuter patterns, traffic forecasts and how the land-use influences the travel demand. GIS-AI-based models have been invoked to forecast the modal shift when infrastructure is being changed or changing policies that make cities more resilient (Aslam et al., 2025). Some of the models that are commonly used are Long Short-Term Memory (LSTM) networks in temporary traffic prediction, Convolutional Neural Networks (CNNs) in recognizing spatial patterns and hybrid ANN-GIS models in multimodal transport prediction. These models are used to complement GIS, since they discern nonlinear relationships among traffic density, land use, and variations in demand over time, and, thus, enhance policy simulation and optimization of the system.

Singh et al. (2020) in India have introduced an AI-enhanced GIS model to examine the traffic flows in the medium cities, and the planners have the chance of simulating the demand in the future under various developmental conditions. This application of ANN modelling to forecast the information obtained on mobile makes their solution in ANN-GIS environment framework more promising to address the issues of urban complexity.

Cloud Platforms, Big Data, and Digital Twins

Volume and velocity of transportation information have necessitated the use of elastic computing environments to execute GIS applications. Cloud based GIS solutions offer this backbone to offer storage, management and analysis of large heterogeneous data in near real time. The portals enable agencies to cooperate in designing, interactive representation of maps as well as access to high-performance spatial modelling computer systems (Batty et al., 2012).

In the case of GIS, big data analytics can be applied to macro-trends around mobility and decisions on micro-levels. Be it analysing the boarding patterns of passengers in the event of using a public transport method of measuring how customers were utilizing parking facilities and the level of foot traffic, when supplemented with GIS, big data will allow one to convert an analysis into actionable intelligence, which can be implemented in real-time to optimize their systems and make future infrastructure investments.

The virtual realization of the urban spaces that mirror real-life urban systems of infrastructure and movement in real-time is one of the revolutionary effects of this integration: the digital twins, the virtual reflection of the urban environment. Planners can also utilize digital twins that are trained on sensor data and can be controlled via predictive algorithms and executed with GIS estimation to deliver the proposed hypothetical scenario and equations of vulnerability and observed the test interventions without implementing any of it to the ground (Tao et al., 2019).

Digital twins are already used in such cities as Rotterdam or Helsinki to model the impact of installing new bike lanes, congestion pricing mechanisms, and intelligent streets redesigns. By using these platforms, planners can visualize causal spaces and time and therefore facilitate them in making superior and more responsive inclusive mobility plans.

GLOBAL CASE STUDIES

Singapore: Real-Time Bus Tracking

Singapore may be viewed as a global model in the application of GIS-based systems to ensure that its mass transport system is effective. LTA has been able to use GIS in bringing together its activities of operating the public buses and having centralized and real time monitoring of the fleet and paths. The GPS-equipped devices installed to buses as well as the data being added to a GIS platform will allow the city-state to monitor the performance of the transit, adjust the timetables, and react dynamically to the disruptions of the services.

The Passenger Information System (PIS) is one of its pillars that give information on the arrival time of the bus in real time using mobile applications and the digital screen at the bus stop. These services can be provided with the help of route mapping and vehicle location analytics through the GIS which makes it more reliable and enables passenger satisfaction (Kumar, 2024b). Moreover, the GIS structure contains predictive algorithms to calculate arrival times and head ways to optimize it and assist in the reduction of congestion and waiting time.

The Intelligent Transport System (ITS) of the LTA includes such applications as GIS-based planning of congestion management schemes and optimization of different routes to signal traffic lights in priority order giving buses and emergency vehicles a route through. The connection of GIS with AI systems and sensors will enable Singapore to have an efficient, flexible, and responsive bus system and sensitive to customer demands and traffic (Lee and Yoon, 2022).

Amsterdam: GIS for Multimodal Bike-Transit Planning

The Amsterdam urban mobility vision maximizes the multimodal integration with consideration on the non-motorized transport. GIS in the city is also very usual since they have combined their cycling tracks with the road and other transport streets such as trams, buses, and metro. Amsterdam facilitates the mode shift by mapping the cycling flows and the parking demand and access to the transit stations spatially, making it trouble-free.

Amsterdam Smart Mobility plan involves the application of GIS to recognize and redesign street space and encourage the best mix of bicycles and transit. The models of accessibility can be visualized in GIS where the shortest and safest routes between the residential zones and the significant transport nodes can be evaluated, and the speed of the cyclist in the heat maps can be estimated to assist in locating the best point where the bike-sharing docks can be placed, and the lanes are secure (Pojani & Stead, 2016).

In addition, GIS is also involved in the Mobility-as-a-Service (MaaS) project which involves route schedule, booking and charging services within transport networks in a single mobile application of the city to commuters. Based on the geospatial insights on the platforms of data analytics, it is possible to forecast travel patterns, define the modality choice, and vary services seasonally or in the event of major events (Pangbourne et al., 2020).

As an example of how in-built GIS can aid its cycling-transit integration policies to allow development of user-centric, sustainable and highly interconnected mobility ecosystem, Amsterdam can be cited. In the same vein, Bogota and Nairobi are examples of GIS application in fast growing cities. GIS-based planning of Bus Rapid Transit (BRT) has optimally served to cover routes in the peripheral communities and has led to a decrease in travel times of these communities in Bogota, and the use of open-source GIS by Nairobi under the Digital Matatus project has mapped informal transit networks, turning data-scanty systems into structured mobility databases that are used to drive urban transport planning and equity-oriented policy-making (Coetzee et al., 2020).

India: GIS-Enabled Route Rationalization

India has seen urbanization of a significant urban growth with population growing at a very rapid rate; therefore, creating problems in the capacity of the transportation system in these cities; particularly those of tier II and III cities. GIS has gained significance in optimization of the transit services and bus route rationalization. Different cities of India have also initiated GIS-based projects to determine their coverage efficacy not to mention that they have removed redundancies to optimize the frequency of the services extended to the commuters in terms of spatial equity requirements.

To give an example, GIS helped the transit authorities in Bhopal and Pune to examine spatial profile of routes, bus stops and bus ridership patterns. Based on this, the maps of population density, types of land-use, and traffic congestion were overlaid on these data layers to determine the service accessibility index and underserved territories (Yenisetty and Bahadure, 2021). The resultant effect was line duplication scrapping, development of feeder networks and expresses inception.

GIS has equally been influential in the procedure of integrating the informal and formal transport services. Wholesale buses and metro networks have been superimposed on the GIS platform and have been digitized in such cities as Delhi and Ahmedabad, and the common routes of auto-rickshaw and minibuses have been overlaid. Integration assists in the realization of multimodal journey planning, single fare system, and real time service monitoring.

The potential of Grade Intersection planning based on GIS in the city of Indore was analysed in a research study by Singh and Khubani (2018) since spatial intelligence was utilized to arrive at the congestion abatement and intersection design solutions to improve the situation with traffic and pedestrian safety.

USA: Open-Source GIS Adoption and Community-Led Mapping

The US has already embraced the open-source GIS systems, and they have helped to engage the people in the transport planning. Other initiatives include OpenStreetMap (OSM) and Federal Highway Administration (FHWA) Geospatial Data Collaboration (GDC) which not only promote the sharing of geospatial data (which is made freely available by anyone) but also community-based mapping and infrastructure planning.

The application of GIS tools in other cities like Portland Oregon has been used to enhance development of active transportation plans. By the combination of the information concerning the GPS applications, citizen surveys and sensors, the urban planners could create all-purpose maps of areas where the bicycle lanes were required, walkway areas needed and transit access gaps. Such insights have informed of the investments in the infrastructure, as well as policy provisions that have been focused on the local level (Goodchild & Glennon, 2010).

San Francisco Municipal Transportation Agency (SFMTA) also must deal with GIS as the indicators of equity such as the volume of income, disability rate, and dependency on transit are also described with the services coverage. The spatial analytics also enable the agency to prepare general transport measures, which would be inclusive and capable of tracing the long-term changes in equity (Karner et al., 2020).

The difference between the U.S. approach and the rest is a decentralized and democratized GIS. Local governments, civic organizations and citizens themselves use and promote open GIS platforms, which helps to achieve a sense of ownership and results in the appearance of innovations in transport planning.

Table 2. Summary of Global Case Studies on GIS-Integrated Urban Mobility

City/Country	Key GIS Application	Innovations Introduced	Impact on Mobility	Primary GIS Functions / Data Types Used
Singapore	Real-time bus tracking	AI-enhanced route optimization	Increased reliability and commuter satisfaction	GPS and IoT-based fleet monitoring, spatiotemporal analytics
Amsterdam	Bike-transit multimodal integration	GIS-based smart cycling infrastructure	Promoted sustainable, active transport	Land-use mapping, multimodal connectivity modelling
India (Pune, Delhi)	Route rationalization, accessibility audits	Integration of informal/formal systems	Improved service coverage and equity	Accessibility indexing, demographic overlays
USA (Portland, SF)	Open-source GIS for community planning	Citizen-led data collection and analysis	Inclusive and transparent mobility planning	Open data platforms, participatory GIS
Bogotá / Nairobi	BRT and informal transit mapping	Open-source GIS integration	Enhanced network equity and data visibility	Route digitization, participatory mapping

Table 2 consolidates insights from global case studies that demonstrate the innovative and localized use of GIS in urban mobility planning.

CHALLENGES AND OPPORTUNITIES

Data Privacy and Interoperability Issues

The wide application of Geographic Information Systems (GIS) in transport planning has brought up new concerns to deal with data confidentiality, safety and interoperability. Since GIS systems are starting to comprise of real-time information regarding GPS, mobile applications, IoT devices and surveillance tools, ethical and legal issues regarding how much and how much personal mobility information is being collected are a lot.

The issue of data privacy is extremely sensitive when it comes to scenarios whereby location-related data can be utilized to track the habits or trends of people or their socio-economic backgrounds. Such data capture could be a contravention of an individual right and abused or discriminatively used without appropriate anonymization policies and consent procedures (Batty, 2019). In many developing countries, especially in India, the legal and policy coverage of geospatial information security is either poorly formulated or unevenly applied and therefore, adds to the setbacks of individuals having faith in smart mobility systems. There is an increasing influence of ethical data governance models that govern transport data management. The location-based data processing standards of anonymization, consent, and accountability presented by the General Data Protection Regulation (GDPR) of the EU and the Digital Personal Data Protection (DPDP) Act 2023 of India serve as the standards to provide GIS-driven transport analytics functions in open and rights-driven digital ecosystems.

The other challenge, and the most crucial one, is the issue of interoperability, particularly in cases when GIS platforms are required to get data from other sources (public transport agencies, municipal bodies, and private companies, application of third party). There are fragmentation and inefficiencies occurring because of difference in data formats, standards and update frequencies. Finally, the ubiquitous APIs or standardized spatial macros does not exist and this kills the cross-platform integration as well as the usefulness of GIS to act as a single source of decision support (Arampatzis et al., 2003). For instance, all the conflicts mentioned above aside, open data can be provided by the metro authorities on one of the standards, but bus companies or traffic control divisions can use their own proprietary standards, and an on-time collaboration and visualization becomes tough.

The issue of interoperability will no longer be simple since the GIS systems become more complex and more networked, raising both technological resolutions by means of open standards as well as institutional intervention by means of data-sharing agreements and metadata standards that govern the same between the stakeholders.

Institutional and Capacity Limitations

It is an established fact that GIS technologies are becoming more easily available; however, institutional readiness to purchase and effectively utilize them is not in the right proportions. Many urban and transport planning departments particularly in the lower and middle-income countries were poor in terms of poor technical infrastructure, lack of skilled workforce, and poor governance structure. These limitations restrict GIS opportunities and delay the achievement of the smart mobility goals.

The most common issue is that there are very few professionals who practice under both the fields of geospatial science as well as transport planning. Spatial analysis, data science, policy evaluation, and infrastructure engineering skills are also required in a cross-disciplinary approach to GIS-based mobility planning, which is not the mainstream in most urban development agencies now (Chen and Li, 2024). GIS will keep underperforming or malfunctioning unless it invests in education continuously and expands professionally.

Institutional silos also restrict the incorporation of the use of GIS. Most cities have traffic controls, modes of transport, land-use policies and environmental surveillance, which are controlled by independent bodies with no or minimal exchange of information and coordination. This silo has resulted in waste of resources being deployed in the same locality and results in the diversity of spatial data that dilutes the validity and quality of planning outputs in GIS (Gijre & Gupta, 2020).

The examples of these problems are the article by Singh and Khubani (2018) on the analysis of traffic in the city of Indore where the absence of coordination in the context of data collection in different modes of transport and between the agencies involved led to the gaps in mitigating the congestion and route planning. According to the outcomes of their work, data governance systems and cross-agency cooperation are critical factors in generating all the possibilities of GIS. Several international collaborations depict developments in capacity building over institutional capacity gaps. The Esri-UN-Habitat partnership has taught sustainable mobility planning using GIS in more than 30 African and Asian cities, and India has already assembled state-level spatial data infrastructure units, which is increasing professionalism and expediting the adoption of GIS-based urban governance projects.

Opportunities from Open Data, Upskilling, and Funding

Despite all these challenges, the opportunities are rising, and they can be highly improved when it concerns the role of GIS applied in transport planning. The prevailing open data trend that is rocking the world evened the scorecard about access to high quality spatial data that could be used by cities, developers and researchers to develop customized GIS applications without going through the exorbitant cost of data ownership and access. The development of locally specific transport models and solutions in mobility has been possible through such open-source projects as OpenStreetMap (OSM), Global Human Settlement Layer (GHSL), or Transport Data Hub of the World Bank (Coetzee et al., 2020).

The other positive trend is the increase in the capacity-building initiatives. GIS and transport modelling courses and certification programs are becoming increasingly accessible in universities, online learning providers and websites of international organizations. They are attempting to fill this skills gap, by training the generational successors of urban planners, data scientists and civil engineers to become trained in geospatial intelligence. In specific cases, the programs of the UN-Habitat together with EsRI provide specific GIS training to the local municipal employees of the developing world and can also be linked with some urban projects.

GIS funding and structure development are also on the rise. The digital spatial systems are now deemed as a strategic tool to governments and multilateral agencies to execute Sustainable Development Goals (SDGs) with specific emphasis on SDG 11 (Sustainable Cities and Communities). Because of this, there are investments in GIS that include smart city programs, digital twins' programmes as well as climate-resilient transport planning.

In India, the Smart Cities Mission and AMRUT program (Atal Mission for Rejuvenation and Urban Transformation) are already beginning to make budgetary allocations of GIS-based applications, in particular, urban mobility, water management, and infrastructure audit (Singh & Khubani, 2018). These kinds of investments are slowly creating an environment where spatial intelligence will be viable in the world of decision-making in the field of the public sector.

In advance, despite technical, institutional and ethical barriers, the climate is far more favourable to the further proliferation of GIS-based transport systems because of free partnerships, specialized training, and eco sustainable funds.

STRATEGIC FRAMEWORK FOR IMPLEMENTATION

Policy Roadmap and Investment Priorities

A rational narrative of the policy encompassing a long-term vision, institutional commitment and budgetary investment should be provided to the decision maker to have realistic results in the scale implementation of GIS-based mobility solutions. The effectiveness of such initiatives is only applicable as far as national, regional and local governments can align their spatial data policies with transportation objectives and can do these in an inclusive, interoperable and sustainable way.

The process of strategic policy framework of GIS integration initiation should begin with the official shift towards spatial intelligence as the powerful element of urban and transport planning. This recognition should then be incorporated into law, geospatial data interoperability national standards, and implemented geospatial data use in any transport-related proposal (Mutuku et al., 2019). Stated as an example, in the European Union, the INSPIRE Directive has enabled a legal framework that allows sharing of geospatial information across sectors with the outcome that cross-border transportation collaboration can be made easy.

The investment must be biased towards building and sustaining GIS infrastructure like data acquisition, cloud storage, a combination of sensors, software tools as well as talented human resources. The budgets of the departments in the government should also allocate funds towards training, stakeholder engagement as also research and development; hardware and software are not the only areas that need funds. The emancipation made on the application of geospatial data and the incorporation of the business sector have positive dynamics through the National Geospatial Policy of India (2022).

It is also necessary to make certain investments in pilot projects and innovation malls that will demonstrate the applicability of GIS in the real-world situation of traffic. Such demonstration projects help in the formation of the proofs, institutional trust and justify future investments. Such core applications of GIS related to traffic clearance, asset monitoring and infrastructure checking are among the main applications of GIS in most smart city endeavours including the one under the Smart Cities Mission in India which suggests the potential of spatial technologies in the context of transportation in the city (Sharma et al., 2021).

Moreover, partnerships with academic institutions and other companies in the private sector can help the company to transfer technology and engage in collaborative innovation. Governments ought to reward start-ups and research centres to develop domain related GIS applications that can be utilized in addressing the regional mobility related challenges such as the integration of rural and urban areas, the last mile connectivity of transportation, or flood resistant transportation design.

Inter-Agency Collaboration and M&E Indicators

Mobility plans based on GIS require institutional cooperation between the various levels of governance to be successful. The urban transport systems are typified by multi-jurisdiction in which road and transport services, development planning, environment and emergency management might be involved. In the absence of an agency cooperation, the isolation of spatial data is realized, the futile work is maintained, and the over-view information is separated.

Checking these hurdles can be done by ensuring that the structures of governance are characterized by data-sharing rates, interoperable transport-planning platforms and steering committees comprising of various stakeholders who can guide the process of transport planning with the help of GIS. The datasets of different agencies can be aggregated via combined geospatial databanks such as urban transport observatories or mobility -data hubs to make tandem analysis and harmonization of policies (Geurs et al., 2008).

Helsinki and Barcelona cities provide good examples of the transport agencies working together with other divisions of the environmental, planning and ICT departments to share common GIS dashboards and digital twins to dynamically control the city. Such examples include UMTA model around the country and the stakeholders of transport in the metropolis proposed the model of unification of planning and budgeting of transport.

The M&E structures should also be geospatialized to be more geospatial indicators. Spatial-related measures, e.g. accessibility index, spatial equity scores, congestion heat maps, etc. should be added to the classical mobility-related metrics, such as VKT, mode shares, or average trip duration. These indicators are available in GIS-enabled format, i.e. they may be traced in real-time, thereby rendering the mobility systems more receptive to changes and more open (R. Liu et al., 2018).

J. Singh et al. (2022) provide the reasons why the performance measures which are spatially disaggregated are highly critical in determining the outcome of smart mobility interventions in Indian cities. Their work underscores the fact that GIS has the potential of changing the disconnection that has been in existence between data filled systems and evidence-based decision making.

In conclusion, therefore, a good policy, institutional integration, and spatially intelligent performance tracking should form the basis of an effective strategic framework to plan the implementation of GIS in the transport planning. Building geospatial thinking into the governance system would allow cities to create more responsive, inclusive and efficient transport networks.

Governance and Participatory Dimensions in GeoSmart Mobility

The effectiveness of GIS-based intelligent mobility systems requires both technology and sound governance and participation systems. With proper governance, geospatial information is disseminated among transport authorities, municipal agencies, and the private mobility providers and institutional silos which tend to hamper integrated planning are broken. The development of open data standards and interoperable platforms can enable various stakeholders to work together in creating transport networks that are efficient, equitable and climate resilient.

The participatory aspect is equally important, whereby GIS is an avenue used in involving communities in planning mobility. By engaging citizens on interactive mapping platforms, they can give feedback on traffic jams, unsafe walking spots, or how new routes can be introduced to their area of residence through the new transport means. This bottom-up feedback allows for improving inclusivity and increasing the trust of people in transport policies. In addition, the use of participatory GIS (PGIS) methods empowers vulnerable populations, including older population or low-income commuters, to shape the decision regarding accessibility and coverage of services. Participatory GIS (PGIS) practices in the reality capture the strength of participatory spatial planning by citizens. In Kenya, the Digital Matatus program used community mapping projects that allowed residents to map the informal routes of transit to enhance the availability of accessibility information to the community. Equally in Brazil, in Porto Alegre, PEinvestigou used PGIS to map pedestrian safety and transport barriers in areas with low incomes, which directly influenced the planning of the municipal mobility. The participatory mapping has been used in the Philippines to assist in the identification of underserved commuting corridors within the jeepney network in Manila to facilitate fair transport reforms.

Participatory GIS forms of governance are transparent, accountable and responsive in decision-making. Spatial intelligence combined with citizen engagement can work together to design mobility systems that can respond to local demands and support the wider sustainability and resilience goals. By so doing, GeoSmart Mobility will not only be a technological intervention but a socially based change.

CONCLUSION AND FUTURE DIRECTIONS

GIS and the Path Toward Autonomous and Resilient Mobility

Geographic Information System (GIS) is already an emerging trend that is fast evolving and becoming the foundation of how the future of urban mobility will be based on instead of being an inactive cartographic system. As cities worldwide strive to achieve autonomous, connected, electric and shared (ACES) transportation, GIS is among the initial technologies that enable one to navigate the space cognizant, decision making and responsive in real-time. One of the autonomous vehicles (AVs) is subject to the accessibility of high-definition maps that have GIS data layers so that it can locate the lanes and detect road signs and adapt to road conditions (Seif & Hu, 2016).

The concept of resilience in transport systems planning is a growing issue that is within the capacity of predicting disruption and reconstruction of systems along the way. GIS is more resilient to such events by analysing risks, impact, and dispatching resources so that it will endure such events during any occurrences of disaster, climatic variations, or socially and politically shocks. GIS mobility solutions are shifting cities to preparedness and ability to recover by evacuating people during floods, reconfiguring a supply chain during pandemics and so on (Saunders and Da Silva, 2009).

Real-time spatial decision-making will form a business norm in the future as the fusion of GIS and sensor networks, vehicle-to-everything (V2X) communication, and edge computing is achieved. This change commits to maintaining innovation and investment in spatial infrastructures of mobility systems to prove them.

Urban Digital Twins, AR/VR Use Cases

The introduction of urban digital twins will be drastic move towards the integration of GIS in immersive planning. Digital twins are used to simulate the real-time of the physical and functional character of the urban systems that allow planning, managing, and imagining citizens and the environment of the complex dynamics of the infrastructures (Batty, 2018). They are GIS-powered in their core, as the IoT sensors, satellite images, transport systems, and land use datasets are integrated. Figure 4 creates a digital image of a digital twin of an urban mobility system, which is driven by GIS and real-time data layers, which allow immersive planning and responsive governance.

The technologies of Virtual Reality (VR) and Augmented Reality (AR) bring the application of the GIS to the field of participation and experience. AR/VR interface can offer the urban stakeholders the option of interacting with the spatial plans, assessing the alternative transport conditions, and posting comments during the stages of the project design. To illustrate the opportunities of GIS, VR, Swedish planners have described the possibilities of replicating the behaviour of pedestrians and wayfinding in redesigns of the streetscapes in Stockholm (Thakuriah et al., 2016).

Figure 4. Urban Digital Twin for Transport Planning (concept)

That way, Figure 4 does not only render the concept of the digital twin visual but also represents the multiplied convergence of technologies previously mentioned, namely, IoT, AI-enhanced GIS, and the integration of real-time information sources in a single platform of visualisation. This connection supports the fact that immersive technologies such as AR and VR enhance the analysis capacity of GIS so that stakeholders can engage with spatial information to promote collaborative and proactive transport governance.

There are technologies that will eventually become more accessible and will be applicable in the inclusive decision-making processes, testing various scenarios, and developing mobility solutions collectively. Advocacy of the use of such tools in the city governance process will lead to the democratization of spatial knowledge and will also increase the consistency of infrastructural investments and the requirements of the communities.

Final Reflection on the Value of Spatial Data in Smart Cities

This chapter has discussed how GIS technologies are changing urban mobility through the supply of geospatial intelligence to integrate them with smart technologies and realize their strategic application in transportation modes and geography. It can be the streamlining of traffic flows and rationalization of the logistics of real transport, or the optimization of current-day digital dashboards and digital twinning, but GIS is one of the versatile and unreplaceable sets of tools in the transport planning landscape of the 21st century.

It will be a technical path, also institutional, where the construction of spatial data infrastructures, the creation of inter-agency collaboration and ethical management of location data are necessary. Such cities, which consider GIS as a strategic resource, but not a technical means of accomplishment, will be more likely to achieve long term, inclusive and resilient mobility.

As the smart city paradigm gains maturity, the foundation of the new generation planning practice will be spatial intelligence. If there is appropriate stewardship, GIS is most likely going to remain a central figure in informing the complexity of urban mobility, and in answering the human-oriented transport futures. The chapter, in totality, provides both conceptual and operation roadmap of institutionalizing GIS-based urban mobility governance by researchers and policymakers. It promotes the convergence of spatial intelligence, digital twins, participatory data systems, and ethical governance to inform a resilient, equitable, and data-driven transport policy. This model can be used as an imitable model by cities that want to streamline smart mobility planning towards sustainability and inclusivity goals in the face of transforming digital and climate requirements.

REFERENCES

Aljoufie, M. (2013). Toward integrated land use and transport planning in fast-growing cities: The case of Jeddah, Saudi Arabia. *Habitat International*, *41*, 205–215. DOI: 10.1016/j.habitatint.2013.08.010

Alkaissi, Z. A. (2024). Traffic congestion evaluation of urban streets based on fuzzy inference system and GIS application. *Ain Shams Engineering Journal*, *15*(6), 102725. DOI: 10.1016/j.asej.2024.102725

Alsubai, S., Dutta, A. K., & Sait, A. R. W. (2024). Hybrid deep learning-based traffic congestion control in IoT environment using enhanced arithmetic optimization technique. *Alexandria Engineering Journal*, *105*, 331–340. DOI: 10.1016/j.aej.2024.06.072

Arampatzis, G., Kiranoudis, C., Scaloubacas, P., & Assimacopoulos, D. (2003). A GIS-based decision support system for planning urban transportation policies. *European Journal of Operational Research*, *152*(2), 465–475. DOI: 10.1016/S0377-2217(03)00037-7

Aslam, M. M., Shafik, W., Hidayatullah, A. F., Kalinaki, K., Gul, H., Zakari, R. Y., & Tufail, A. (2025). Intelligent Transportation Systems: A Critical review of integration of Cyber-physical Systems (CPS) and Industry 4.0. *Digital Communications and Networks*. Advance online publication. DOI: 10.1016/j.dcan.2025.06.014

Bachechi, C., Po, L., & Rollo, F. (2021). Big data analytics and visualization in traffic monitoring. *Big Data Research*, *27*, 100292. DOI: 10.1016/j.bdr.2021.100292

Batty, M. (2018). Digital twins. *Environment and Planning. B, Urban Analytics and City Science*, *45*(5), 817–820. DOI: 10.1177/2399808318796416

Batty, M. (2019). Urban analytics defined. *Environment and Planning. B, Urban Analytics and City Science*, *46*(3), 403–405. DOI: 10.1177/2399808319839494

Batty, M., Axhausen, K. W., Giannotti, F., Pozdnoukhov, A., Bazzani, A., Wachowicz, M., Ouzounis, G., & Portugali, Y. (2012). Smart cities of the future. *The European Physical Journal. Special Topics*, *214*(1), 481–518. DOI: 10.1140/epjst/e2012-01703-3

Chen, T., & Li, Y. (2024). Building urban resilience: Lessons from the COVID-19 pandemic for future-proofing city infrastructure. Journal of Urban Management. https://doi.org/DOI: 10.1016/j.jum.2024.11.016

Coetzee, S., Ivánová, I., Mitasova, H., & Brovelli, M. (2020). Open Geospatial Software and Data: A Review of the Current State and A Perspective into the Future. *ISPRS International Journal of Geo-Information*, *9*(2), 90. DOI: 10.3390/ijgi9020090

Delbosc, A., & Currie, G. (2011). Using Lorenz curves to assess public transport equity. *Journal of Transport Geography*, *19*(6), 1252–1259. DOI: 10.1016/j.jtrangeo.2011.02.008

Du, H. (2024). Urban Landscape Accessibility Evaluation Model based on GIS and spatial analysis. *Results in Engineering*, *103723*. Advance online publication. DOI: 10.1016/j.rineng.2024.103723

Elgayar, A., & Jrade, A. (2022). Integrating 3D modeling and GIS-Based land use and Multimodal Transport Accessibility Evaluation model to design and retrofit sustainable bridges. In Lecture notes in civil engineering (pp. 83–93). https://doi.org/DOI: 10.1007/978-981-19-0968-9_7

Geurs, K. T., Boon, W., & Van Wee, B. (2008). Social Impacts of Transport: Literature review and the state of the practice of transport appraisal in the Netherlands and the United Kingdom. *Transport Reviews*, *29*(1), 69–90. DOI: 10.1080/01441640802130490

Gijre, V., & Gupta, S. (2020). Urban transport governance practice and challenges in an Emerging Economy – Case study of India. *Transportation Research Procedia*, *48*, 2435–2445. DOI: 10.1016/j.trpro.2020.08.293

Goodchild, M. F., & Glennon, J. A. (2010). Crowdsourcing geographic information for disaster response: A research frontier. *International Journal of Digital Earth*, *3*(3), 231–241. DOI: 10.1080/17538941003759255

Huang, X., Wang, S., Lu, T., Liu, Y., & Serrano-Estrada, L. (2024). Crowdsourced geospatial data is reshaping urban sciences. *International Journal of Applied Earth Observation and Geoinformation*, *127*, 103687. DOI: 10.1016/j.jag.2024.103687

Karner, A., London, J., Rowangould, D., & Manaugh, K. (2020). From transportation equity to transportation justice: Within, through, and beyond the state. *Journal of Planning Literature*, *35*(4), 440–459. DOI: 10.1177/0885412220927691

Kumar, S. (2024b, December 18). Singapore Smart Traffic System is Redefining Urban Mobility. IntelliStride. https://www.intellistride.com/blog/how-singapores-smart-traffic-system-is-redefining-urban-mobility/

Lee, C., & Yoon, Y. (2022). A novel bus arrival time prediction method based on Spatio-Temporal flow centrality analysis and deep learning. *Electronics (Basel)*, *11*(12), 1875. DOI: 10.3390/electronics11121875

Liang, Y., Li, J., & Liu, H. (2021). GIS-based analysis of urban traffic model: Case study in Beijing. E3S Web of Conferences, 268, 01056. https://doi.org/DOI: 10.1051/e3sconf/202126801056

Liu, R., Chen, Y., Wu, J., Xu, T., Gao, L., & Zhao, X. (2018). Mapping spatial accessibility of public transportation network in an urban area – A case study of Shanghai Hongqiao Transportation Hub. *Transportation Research Part D, Transport and Environment*, *59*, 478–495. DOI: 10.1016/j.trd.2018.01.003

Liu, X., Payakkamas, P., Dijk, M., & De Kraker, J. (2023). GIS Models for Sustainable Urban Mobility Planning: Current use, future needs and Potentials. *Future Transportation*, *3*(1), 384–402. DOI: 10.3390/futuretransp3010023

Mutuku, B., Boerboom, L., & Madureira, A. M. (2019). The role of Planning Support Systems in national policy transfer and policy translation in secondary cities. *International Planning Studies*, *24*(3–4), 293–307. DOI: 10.1080/13563475.2019.1657809

National Geospatial Policy. 2022. (2022). https://www.pib.gov.in/PressReleasePage.aspx?PRID=2106569

Pangbourne, K., Mladenović, M. N., Stead, D., & Milakis, D. (2020). Questioning mobility as a service: Unanticipated implications for society and governance. *Transportation Research Part A, Policy and Practice*, *131*, 35–49. DOI: 10.1016/j.tra.2019.09.033

Papa, E., & Bertolini, L. (2015). Accessibility and transit-oriented development in European metropolitan areas. *Journal of Transport Geography*, *47*, 70–83. DOI: 10.1016/j.jtrangeo.2015.07.003

Pojani, D., & Stead, D. (2016). The urban transport crisis in emerging economies: An introduction. *Urban Transport in the Developing World*, 1–10. https://doi.org/DOI: 10.1007/978-3-319-43851-1_1

Saunders, M. J., & Da Silva, A. N. R. (2009). Reducing Urban Transport Energy dependence: A new Urban Development Framework and GIS-Based tool. *International Journal of Sustainable Transportation*, *3*(2), 71–87. DOI: 10.1080/15568310701648037

Schiller, P. L., & Kenworthy, J. R. (2017). An introduction to sustainable transportation. In Routledge eBooks. https://doi.org/DOI: 10.4324/9781315644486

Seif, H. G., & Hu, X. (2016). Autonomous driving in the iCity—HD maps as a key challenge of the automotive industry. *Engineering (Beijing)*, *2*(2), 159–162. DOI: 10.1016/J.ENG.2016.02.010

Sharma, P., Singh, R., & Srivastava, A. (2021). Analyzing the role of geospatial technology in smart city development. In The urban book series (pp. 1–20). https://doi.org/DOI: 10.1007/978-3-030-71945-6_1

Shi, W., Kong, Q., & Liu, Y. (2008). A GPS/GIS Integrated System for Urban Traffic Flow Analysis. In 2008 11th International IEEE Conference on Intelligent Transportation Systems (pp. 844–849). IEEE. https://doi.org/DOI: 10.1109/ITSC.2008.4732569

Singh, J., Chaudhary, H. K., & Malik, A. (2022). A Case Study on Indore BRTS with Reference to Other Indian Cities. *International Journal for Research in Applied Science and Engineering Technology*, *10*(10), 196–203. DOI: 10.22214/ijraset.2022.46976

Singh, S. K., Alasali, B. E., & Belbase, Y. (2025). Urban Transportation planning. In *Advances in computational intelligence and robotics*. (pp. 469–504). *IGI Global Scientific Publsihing*. https://doi.org/DOI: 10.4018/979-8-3693-7250-0.ch017

Singh, S. K., & Bansal, A. (2021). Critical analysis and feasibility of PRT for Gurgaon City. [IJSRSET]. *International Journal of Scientific Research in Science, Engineering and Technology*, *8*(1), 83–86. DOI: 10.32628/IJSRSET21812

Singh, S. K., Jain, R., Palaniappan, D., & Parmar, K.. (2025). *Spatiotemporal analysis and machine learning-based prediction of air quality in Indian urban*. Environmental Research and Technology. DOI: 10.35208/ert.1587308

Singh, S. K., & Khubani, P. (2018). Traffic analysis of grade intersections and measures of congestion mitigation at Indore. *Pyrex Journal of Engineering and Manufacturing Technology*, 3(2), 10–19. https://www.researchgate.net/publication/325381098

Singh, S. K., & Saraswat, A. (2019). Traffic simulation & integration using SUMO Simulator. *14th International Conference on Recent Trends in Engineering, Applied Science & Management (ICRTESM-19)*, 327–332. ISBN: 978-93-87793-71-2

Solé-Ribalta, A., Gómez, S., & Arenas, A. (2016). A model to identify urban traffic congestion hotspots in complex networks. *Royal Society Open Science*, *3*(10), 160098. DOI: 10.1098/rsos.160098 PMID: 27853535

Tao, F., Qi, Q., Wang, L., & Nee, A. (2019). Digital Twins and Cyber–Physical Systems toward Smart Manufacturing and Industry 4.0: Correlation and Comparison. *Engineering (Beijing)*, *5*(4), 653–661. DOI: 10.1016/j.eng.2019.01.014

Thakuriah, P., Tilahun, N., & Zellner, M. (2016). Seeing cities through big data. In Springer geography. https://doi.org/DOI: 10.1007/978-3-319-40902-3

Xhafa, S., & Kosovrasti, A. (2015). Geographic Information Systems (GIS) in urban planning. *European Journal of Interdisciplinary Studies*, *1*(1), 85. DOI: 10.26417/ejis.v1i1.p85-92

Yenisetty, P. T., & Bahadure, P. (2021). Assessing accessibility to ASFs from bus stops using distance measures: Case of two Indian cities. *Land Use Policy*, *108*, 105567. DOI: 10.1016/j.landusepol.2021.105567

Zaroujtaghi, A., Mansourihanis, O., Tayarani, M., Mansouri, F., Hemmati, M., & Soltani, A. (2025). *A Systematic Review of GIS Evolution in Transportation Planning: Towards AI Integration*. MDPI. DOI: 10.3390/futuretransp5030097

Chapter 6
Regulatory and Zoning Applications

S. Ida Evangeline
https://orcid.org/0000-0003-2997-7897
Government College of Engineering, Tirunelveli, India

ABSTRACT

This chapter examines the integration of Building Information Modeling (BIM) and Geographic Information Systems (GIS) in the transformation of regulatory and zoning processes, highlighting their potential to enhance precision, transparency, and adaptability in urban governance. Traditional zoning and permitting systems, built on static maps and manual review, have struggled to meet the demands of rapidly evolving, data-rich urban environments. With BIM providing architectural and infrastructural intelligence and GIS giving the spatial, environmental, and socio-economic context, a city can visualise complex regulatory parameters in immersive 3D, implement automatic verification of compliance, and test the effects of the impact of policies on digital twins. The chapter follows the development of zoning patterns, technical approaches to BIM-GIS integration in regulatory visualization and automated permitting, and the analysis of how digital twins can be used in real-time monitoring/testing of scenarios. Legal, institutional and ethical aspects that are considered to enhance and support responsible and equitable adoption include data governance, cross-agency interoperability, and their functionality as publicly accessible information. Best practice case studies by major cities like Singapore, Helsinki, and Rotterdam show the real results that can be achieved in streamlining permitting and public engagements as well as facilitating climate resilient plans. In the future, the chapter outlines the future trends in predictive zoning, IoT-informed adaptive control, blockchain-mediated workflow, and immersive citizen engagement tools. As a tool to bridge the gap between technology and policy, BIM-GIS integration is not just an upgrade to the operational level, but a facilitator of the next

DOI: 10.4018/979-8-3373-4058-6.ch006

generation of spatial intelligence in regulatory and zoning governance-flexible to allow cities to move beyond reactive enforcement to proactive, participatory, and informed decision-making.

1. INTRODUCTION

Regulatory and zoning systems form the backbone of urban governance, shaping the spatial, social, and economic fabric of cities. These frameworks determine where and how development occurs, influencing everything from building heights and land-use patterns to environmental protection and infrastructure provision. Historically, regulatory has been based on stagnant, print-based maps, manual code interpretation, slow review of permits, all of which are probably not able to keep up with the growing speed of modern urban development (Ahmad, 2023).

In recent years, the convergence of Building Information Modeling (BIM) and Geographic Information Systems (GIS) has opened new frontiers for regulatory innovation. BIM offers complex, object-based modelings of both buildings and infrastructure, representing the architectural, structural and operational characteristics of buildings in a rich digital model. GIS instead provides useful tools in locating these designs in wider spatial, ecological and social-economic backgrounds. Together, these technologies provide a new degree of spatial intelligence that is both detailed and contextual: an ability to examine a building in the situation of its environs and the regulatory apps that put them together (Li et al. 2020).

This integration holds major questions of zoning and permitting. Complex zoning codes could be checked automatically against a BIM model, which is geospatially nested in a GIS and, in real time and 3D, tested against a variety of situations within the regulatory framework. The outcome is a more transparent, efficient, and responsive system, a system that can allow the latter as well as compliance enforcement.

The purpose of this chapter is to examine how BIM-GIS integration is reshaping the landscape of regulatory and zoning applications. It will explore the historical foundations of zoning systems, the technical underpinnings of BIM-GIS interoperability, and the ways in which these tools are being leveraged to modernize compliance workflows, enable digital twin-based policy testing, and foster collaborative, data-driven governance. By bridging technology and policy, this chapter positions BIM-GIS integration as a catalyst for smarter, fairer, and more sustainable urban development.

Figure 1. Conceptual Diagram of BIM-GIS Integration.

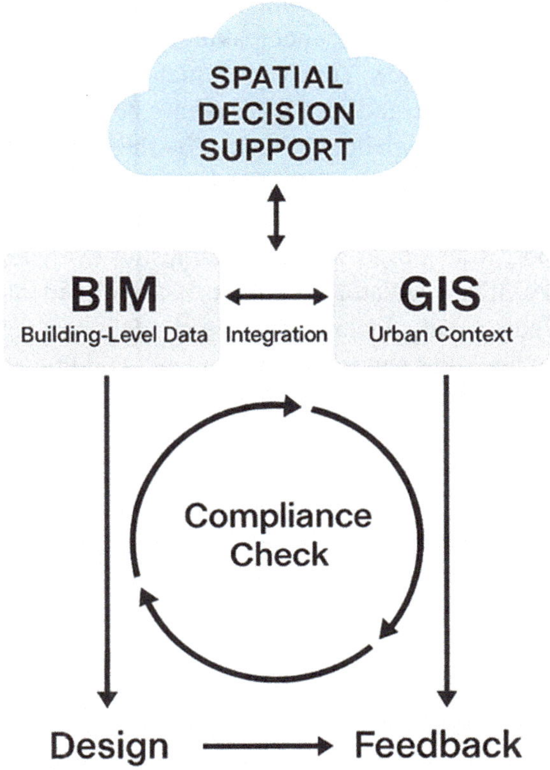

Figure 1 illustrates the integration between Building Information Modeling (BIM) and Geographic Information Systems (GIS), which together form the core of spatial decision support in regulatory workflows. BIM provides detailed, object-level data about buildings and infrastructure, while GIS offers broader contextual layers such as environmental and socio-economic factors. When integrated, these systems enable real-time compliance checking, iterative design-feedback loops, and data-driven urban governance. The diagram emphasizes how this cyclical process supports proactive and adaptive regulatory practices.

2. REGULATORY AND ZONING FRAMEWORKS: CURRENT STATE AND CHALLENGES

Zoning and regulatory systems have long served as the structural framework within which urban development takes place. Originating in the early 20th century, zoning was initially conceived as a means to organize land use, prevent incompatible activities from occurring in close proximity, and protect public health, safety, and welfare. Initial (and, at that time, groundbreaking) examples include the landmark 1916 Zoning Resolution of New York City, which pioneered land-use typology, restrictions of building height, and profiles, and have provided a model of approach worldwide throughout municipalities. Eventually, these models yielded to more complex ones, covering everything related to land-use demarcation to much broader set of issues that are related to cities such as environmental protection, infrastructural capabilities, transport strategies as well as social justice (Vergara-Perucich, 2025).

In contemporary cities, regulatory structures have evolved into complex, multi-layered systems. These usually include land-use regulations which determine what may and may not be done on a particular piece of land, building codes which govern building safety and efficiency, environmental laws which govern emissions and resource consumption and planning to encourage mixed-use or affordable housing development. Each of such layers can be managed by various departments, such as planning, environmental regulatory, transport, or heritage conservation agencies that operate using various sets of data and their processes (differing procedural workflows and institutional interests). Such a divided governance situation usually leads to inefficiencies, wastefulness, and an inconsistent application of policy (Martino et al. 2024).

Despite the sophistication of current frameworks, they remain rooted in tools and practices developed in an era of static information. The two-dimensionality of maps conveying the regulations, text on legislation, and table data remains the primary means. Such materials need specialized interpretation making them a roadblock to professionals who want to manage to keep up with the various competing codes, or to citizens who want to know how regulations are affecting their neighborhoods. The process of checking the compliance is still a predominantly manual one based on people reviewing the design submissions against documents that are regulatory in nature. This not only brings in delays in permitting processes but provides ample opportunity to interpret and err subjectively (Lorenzo-Eiroa, 2022).

Moreover, conventional zoning systems are limited in their capacity to integrate real-time spatial, environmental, and socio-economic data. Some project can be reviewed against the compliance with zoning law without the density of external issues such as traffic patterns and flood hazard or population movement into account, which is desirable in effective urban planning. The implementation of policy can be highly reactive: new zoning ordinances are created and implemented prior to being put into realistic conditions and tested. Another issue is noticed by public engagement which is usually mediated with the use of technical reports, legal messages as well as hearings which often cannot be perceived by ordinary people, thus, lowering transparency and confidence in the regulatory activity (Liu et al. 2025).

Table 1 presents the comparison of traditional vs BIM-GIS enhanced zoning systems. The pressure to address these shortcomings is mounting. Cities are grappling with rising population densities, the urgent demands of climate resilience, housing affordability crises, and the need to transition toward more sustainable patterns of growth. The ability to meet these challenges necessitates more responsive, accurate and integrative regulatory systems. Just as Building Information Modeling (BIM) and geographic information system (GIS) develop, a new avenue to re-imagine zoning and permitting practices becomes possible. BIM-GIS integration has a potentially profound effect on transforming rule-based regulatory processes into dynamic, transparent, and an adaptive tool of governance by digitizing laws into spatially intelligent entities, automating them using computational rules, and facilitating interagency interoperability of data. Such changes precondition the upcoming elaboration of BIM-GIS applications in the following section of this paper.

Table 1. Comparison of Traditional vs BIM-GIS Enhanced Zoning Systems.

Feature	Traditional Zoning System	BIM-GIS Enhanced System
Data Format	2D maps, legal text	3D spatial models, rule engines
Review Process	Manual inspection	Automated compliance checks
Public Engagement	Static reports, hearings	Interactive 3D simulations
Interagency Coordination	Siloed workflows	Integrated data platforms
Adaptability	Reactive	Predictive and dynamic

3. BIM-GIS INTEGRATION FOR ZONING VISUALIZATION AND ANALYSIS

The integration of Building Information Modeling and Geographic Information Systems introduces a transformative dimension to zoning analysis, enabling planners, regulators, and stakeholders to interpret and apply complex rules in ways that are both more precise and more intuitive. BIM models complex, object-based models of buildings and infrastructure, their physical geometry, material properties, and functional characteristics. In contrast, GIS will give the spatial, environment and socio-economic context of these structures. Together, the resulting digital environment will be much more than the sum of its parts: it will enable a proposed development to be compared not only in isolation, but against each of its surrounding environment, applicable regulations, and wider urban systems (Wang & Xie, 2024).

One of the most immediate benefits of this integration lies in the visualization of zoning parameters within a three-dimensional, spatially accurate environment. Instead of using two dimensional zoning maps and purely table-driven zoning codes, instead the stakeholders can use interactive, immersive models, enabling dynamic overlay of zoning lines, height limits, setbacks, and floor area ratio (FAR) limits to a proposed design. These models can show where a building may be cutting off a height envelope, entering a setback zone, or over the allowed density a structure is having and these may not be evident during a lengthy manual check process (Tan et al. 2023).

This capability fundamentally changes the regulatory review process. Instead of parsing through abstract legal descriptions, decision-makers and community members can see in real time how a building fits—or fails to fit—within the regulatory landscape. Such structural compliance visualisations do not have to stop with structural compliance: they can also include environmental limitations, like floodplains, zones of heat islands, heritage overlays, so better picture the consequences of the development. To illustrate this, a BIM-GIS model may show that a proposed building is within an acceptable height and land-use, but within a high-risk flood zone resulting in redesign or other risk mitigation actions being undertaken (Cao et al. 2023).

In addition to static visual assessments, BIM-GIS platforms enable the simulation of multiple design and zoning scenarios. A planners team may be able to model what would happen to the skyline as zoning height limits are adjusted or what might occur to the flow patterns of traffic, sunlight exposure, or access to public amenities as land-use allocations are altered. This set of simulations can be repeated, giving it a design-intent/regulatory-constrained feedback loop. The result does not lay in simple checks of compliance but in hard, informed negotiation between policy objectives and development proposals (Pelden et al. 2025).

Increased transparency and accessibility are also facilitated by the integration. BIM-GIS data can be used to make the interactive 3D models for using in an online engagement platform that allows residents to navigate to see what the proposed development would look and act like in their neighborhoods. This shift to pictorial, data-dense communication can help to fill the gap between the technical domain of regulation and the domain of popular knowledge and awareness thereby enhancing a more-informed civic discourse and of de-conflicted planning decisions.

The visualizations and analysis of zoning that the BIM-GIS combination facilitates are transforming since it goes beyond abstract codes to palpable, spatially grounded experiences. It helps regulators work more accurately, developers build more fully aware of constraints, and communities become involved in the planning process more equally informed. This change also preconditions the very next step of the regulatory innovation the automation of the compliance and permitting which is going to be the consideration of the next part.

4. AUTOMATED COMPLIANCE AND PERMIT PROCESSING

The blending of BIM and GIS leads to much more than merely improving the appearance of zoning restraints; it also introduces the opportunity to automate compliance check and permit issuance processes. Manual and usually piecemeal, the examination of a development proposal against zoning ordinances, and building codes is part of conventional regulatory operations. Not only is it time-consuming, but this also requires cross-checking between design drawings, planning documents, and various regulator texts may have inconsistency in interpretation that is likely to be performed by the planners and inspectors. With the combination of BIM-GIS integration, computer rule engines and artificial intelligence, several of these steps can be automated, lowering the burden on administration and reducing potential errors (Yau et al. 2024).

Figure 2. Automation Workflow for Permitting.

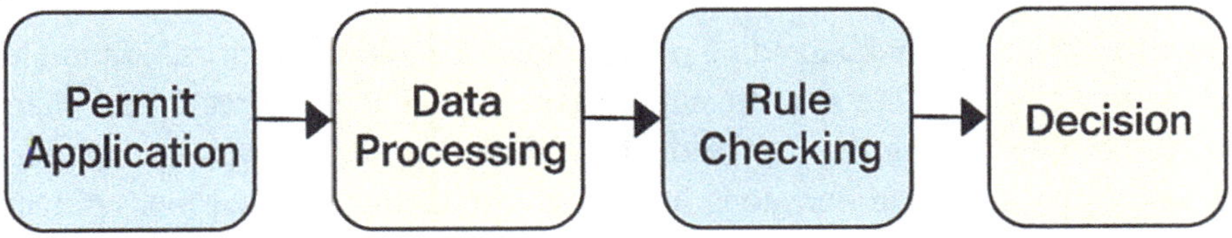

Figure 2 illustrates the automation process in permitting workflows enabled by BIM-GIS integration. The sequence begins with digital permit application submission, followed by backend data processing, automated rule checking against zoning and building codes, and culminates in a system-generated decision. By minimizing manual interventions in routine cases, this workflow reduces delays, enhances regulatory consistency, and allows human reviewers to focus on complex or discretionary matters.

At the heart of this change is the transformation of zoning laws, building codes and environmental needs into machine-readable sets of rules. Against these rules a BIM model, precisely-geolocated in a GIS management environment, can be programmatically tested in real time. As an example, an automated engine could have the capacity to determine whether a proposed building is more than the maximum height permitted in its zone, is already located in a required setback or infringes a land-use requirement. In addition to geometric parameters, the system can be used to measure adherence to performance-based standards, e.g. access to daylight, shadow casting, energy efficiency requirement, or stormwater run off thresholds, all placed into the context of geographic information represented by GIS datasets (Chiodelli & Morpurgo, 2022).

The benefit of such automated checks is that they are quite an asset during design in early stages. Rather than finishing a proposal and then having to wait weeks before it is rejected due to lack of compliance, architects and engineers can get immediate feedback in the environment and quickly change plans before it is too late. Such design-based approach minimizes rework, shortens the process of permitting, and supports more predictable regulated results (Castellanos-Ardila et al. 2021).

Permit application and reviewing are also a part of an effort made by automation. Many of the earlier stages of the review can be algorithmically conducted with digital submission platforms that connect to a BIM-GIS compliance engine prior to a human inspector entering the process. In certain jurisdictions, common permits e.g. to perform trivial repairs or extensions might be approved nearly in real-time once the automated checks indicate full compliance. By adopting this approach, regulatory personnel are released to deal with more advanced or discretionary cases which needs professional judgment (Lindgren, 2020).

Notably, automation does not substitute human control, but it complements it. On the one hand, algorithms do well at examining measurable parameters and less well when it comes to understanding more subtle urban design factors, cultural heritage contexts, or even site-specific circumstances. Planners and regulators change their role to interpreting the results of the automation and answering edge cases and making policy-based decisions instead of conducting manual measurements (Al-Mhdawi et al. 2023).

Transparency and accountability can also be improved by use of BIM-GIS automation combined with the e-government systems. Secure, shared portals ensure that applicants, regulators and the wider population can follow a project in its status as to compliance with the various requirements, and in making clear what has been achieved and what is still pending. In addition to enhancing confidence in the process, this also makes a formal, electronic record, which can be used in making future policy analysis. Given that cities are shifting to the use of digital twins and smart regulatory systems, automation of compliance and permitting is one of the basic steps. BIM-GIS enable systems in preparing the foundation of real-time issues adaptive governance by enforcing zoning laws and building polices in a consistent pattern that is faster and visible (Chong et al. 2025).

5. DIGITAL TWINS AND POLICY SCENARIO SIMULATION

Where automated compliance systems can be used to check proposed developments against the regulations that exist, digital twins would further extend this functionality by establishing constantly evolving, constantly updated virtual replicas of the physical landscape. A digital twin is an interactive 3D model, but it is a living, simulation powered by data that incorporates real-time updates on sensors, environmental monitoring systems, and operations databases. Digital twins are also strong regulatory tools when combined with BIM-GIS integration, allowing planners and policymakers to simulate scenarios, predict the effects, and direct laws even before implementing them in the physical world (Nica et al. 2023).

Figure 3. Digital Twin Policy Simulation Use Cases.

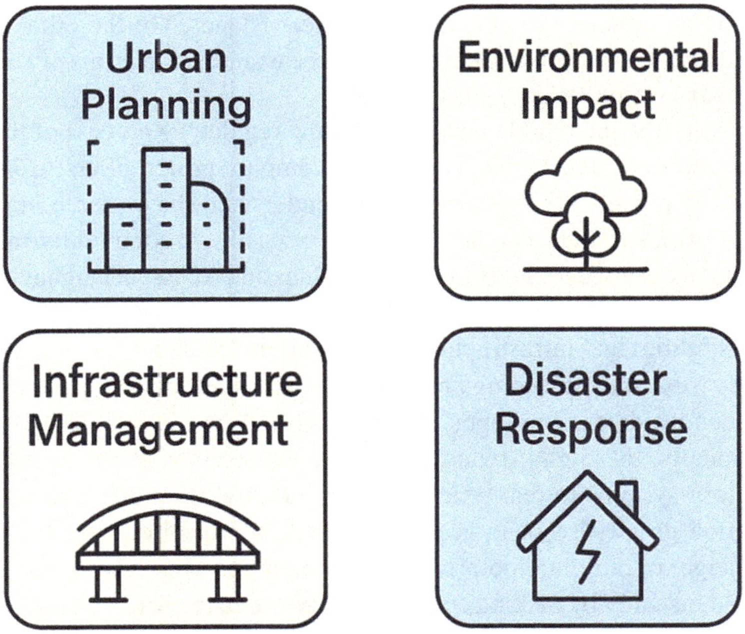

Figure 3 highlights four key applications of digital twins in urban regulatory contexts: (1) urban planning impact analysis, (2) environmental risk forecasting, (3) infrastructure performance monitoring, and (4) disaster preparedness modeling. By linking real-time data to simulated models, digital twins empower policymakers to assess potential outcomes of zoning changes, infrastructure investments, and policy interventions before implementation. These simulations foster proactive, evidence-based urban governance.

In the area of regulation, a BIM-based building model incorporated into a GIS-based digital twin of the city can be applied to model the impact of the zoning change or infrastructure investment that is being proposed. As an example, by modifying a height restriction in a downtown zone of a simulation, one can instantly determine the alterations of skyline, sunlight in the interior, air currents, and shadow effects on the nearest open spaces. On the same note, policy changes involving land-use allocations can be evaluated with regard to its effect on traffic congestion, public demands in transport, and availability of vital services (Hu et al. 2023).

The predictive power of digital twins-a regulatory sandbox- allows policy makers to test possible policy changes in a safe and risk-free environment, to territorial space and time on a realistic domain. Such ability is particularly useful in solving long-range issues of cities, including climate resilience or demographic changes. To give a hypothetical example, an amendment to a zoning code meant to heighten housing density may be tested to determine how the zoning would create a burden on water supply systems, or enhance urban heat islands. On the other hand, the provision of green area requirements may also be examined in terms of cooling and stormwater absorption (Wang, 2021).

There is also the increased engagement in the regulatory process of the general population due to digital twins. Translating complex policy offers to interactive intuitive visual models allow the citizens to understand the possible implications of policy offerings to their neighborhoods more easily. A community meeting no longer has to be centered around largely static plans and technical language; they can plan the proposed development with interactive walk-throughs, and switch between zoning possibilities and infrastructure configurations. Such a level of transparency can enhance trust to the planning processes as well as lead to more healthy discussions between residents, developers and regulators (Wang et al. 2022)

Operationally, the digital twins system excels because of the opportunity to run regulatory systems in real time on top of a digital twin infrastructure. When a construction project is not in accordance with approved designs (i.e., exceeds approved height or building footprint or re-modeling in other ways), the mismatch can be raised instantly by making comparisons between the actual sensor data, BIM models and the digital twin. This will not only enhance the enforcement, but also promote adaptive regulation, which means that policies can be altered dynamically based on observed trends (Wang et al. 2025).

The implementation of digital twins in regulatory and zoning processes is a step towards taking control of governance into much more proactive control of issues, rather than the reactive regimes that many of today policies are, wherein a policy has to wait until a problem has occurred to arrive and control it. Cities will no longer be able to merely impose regulation but they will actively be able to affect the circumstances under which urban life occurs by connecting BIM precision with the spatial intelligence of GIS in a digital environment that is continuously being changed (Sullivan, 2022).

This change towards the anticipatory planning and dynamic policy testing brings up new opportunities, however, there are very important questions regarding data governance, coordination of institutions, and ethical usage, which will be raised in the next section.

6. LEGAL, INSTITUTIONAL, AND ETHICAL CONSIDERATIONS

Utilization of regulatory and zoning processes with BIM-GIS integration and digital twins has pre-eminent potentialities of accuracy, productivity, and inclusivity. Nevertheless, there are also high legal, institutional, and ethical concerns related to these improvements that need to be observed in order to allow the technology to continue establishing a secure urban governance basis. The same advance technologies and tools that have been seen to bring about transparency may by default aggravate opaqueness, discrimination, or un-equal influence due to the lack of premeditated interoperability, data stewardship, and equitable access frameworks. Legally, the main obstacle is determining congruencies between the results of BIM-GIS systems and regulatory procedures that can be enforced with the help of legislature (Martin & Freeland, 2021). The legal tools of zoning ordinances, construction codes and environmental codes are all subject to procedure and are enforceable through due process. In case business is deemed to violate regulations, flagged by an automated compliance engine, the results should be legally defensible in any court of law, demanding that the rules employed, the datasets and ways by which conclusions were arrived at be documented clearly. It increases the necessity of consistent data definitions and computational rule specifications that are technically correct in addition to being consistent with the language of law (Xie et al. 2025).

The institutional considerations focus on governance and interoperability. The regulatory processes frequently cut across several agencies: planning, environmental protection, transportation, utilities each with their own lines of data systems, terms and procedures requirements. These need to be put together into a working BIM-GIS regulatory framework, not just by technical solutions to data exchange but by agreement on governance procedures, allocation of responsibility and common funding mechanisms. Unless there is an institutional coordination, the possibility exists that BIM-GIS platforms will only become isolated tools rather than a hub of complex decision-support systems (de Vries, 2023).

Moral aspects are also decisive. The datasets that are used by BIM-GIS systems in general, and digital twins in particular, may be very large, with a wealth of information about buildings, infrastructure, environmental conditions, and even human activities. This information should be handled with responsibility, through well-developed privacy, data security and informed consent policies. Then there is the issue of algorithmic transparency the logic of AI-enhanced rule engines should be transparent when they are used to run compliance checks it should be accessible to transparently understand that regulatory decisions are not being driven by hidden biases.

Additional ethical mandates are equity and inclusion. Although the state-of-the-art in visualization and simulation may seem like a means to render regulatory processes more transparent to the general-public, it may also have the unintentional consequence of privileging more technical-literate stakeholders or those advantaged by digital connectivity. Ensuring wide accessibility implies creating interfaces and engagement strategies that should be, intuitive, multi-lingual and also accessible by various channels, physical communities spaces included. Also, choices regarding which scenarios to simulate, which data to focus on, how to trade off policy analysis should be not made technocratically but with relevant input in the community (Razinkina et al. 2022).

In a nutshell, the effective incorporation of BIM-GIS into the regulation processes will not be solely based on the technology capacity, but it will also require the establishment of the institutional, legal and ethical scaffolds to shape its application. Preempting these issues will enable cities to take advantage of the accuracy and flexibility of digital technologies without sacrificing concepts of fairness, accountability, and democratic rule.

The next section will turn from these overarching governance considerations to practical, real-world examples, examining how municipalities around the world are implementing BIM-GIS systems for zoning reform, permitting digitization, and smart regulatory frameworks.

Table 2. Case Study Highlights

City	Key BIM-GIS Application	Outcome
Singapore	Automated compliance & Virtual Singapore twin	Faster permitting, inter-agency collaboration
Helsinki	3D city models for citizen engagement	Improved transparency and trust
Rotterdam	Resilience-based planning using flood data	Climate-adaptive zoning strategies
South Korea	National BIM-GIS platform	Scalable compliance automation

7. CASE STUDIES AND BEST PRACTICES

The potential impact of BIM-GIS integration on regulatory and zoning is ceasing to be a theoretical possibility. These technologies allow more nimble governance, permitting and improved zoning transparency and are being deployed in pioneering municipalities across the globe. These case studies will give a good idea of the opportunities and the practical difficulties of instilling the digital conception of space in regulatory regimes. Table 2 presents the highlights of case studies.

Singapore is one of the leading countries to spend on the integration of BIM-GIS in regulatory processes. The city-state has created an end-to-end nation-scale digital twin through its Integrated Digital Delivery (IDD) initiative and the Virtual Singapore project, which encompasses not only the BIM models of buildings, but also high-resolution geospatial data. The twin also allows regulatory authorities to automate compliance checks that confirm that proposed projects do not go against building codes, environmental standards, and zoning standards when the appropriate approvals are issued. The system also facilitates inter-agency coordination thus allowing proposals to be evaluated by planning, transport and environmental departments in one digital space at once (del Cerro Velázquez & Morales Méndez, 2021).

Helsinki, Finland has used the 3D city model in a 3D city model and open data policy to increase transparency and engagement of citizens in the zoning decisions. The city digital twin combines BIM information of developers with land use, transport, and environmental conditions GIS layer data. Proposed development may be simulated in an interactive way so that residents may observe what new buildings would look like integrated into the urban environment casting shadows upon streets and affecting viewshed. The strategy has not only enhanced the capacity of the people to trust the planning process but has also offered a better insight of the cumulative urban effects to the city officials (Gonthier, 2022).

Also, in Rotterdam, the Netherlands, where a complicated relationship exists between the city and water, BIM-GIS coordination has played a key role in the city management. By connecting development proposals to flood risk maps, subsidence monitoring information, and climate adaptation strategies using spatial intelligence, the city can evaluate much more than regulatory compliance: it can evaluate long term resilience. Potential zoning plans are simulated in a digital twin to predict how they would impact a drain, water storage amount, and exit pathways in case of an emergency. Such an active simulation capacity has played a fundamental role in ensuring that the urbanization of Rotterdam is conducted in accordance with its climate adaptation policy (Wu et al. 2026).

Beyond one specific city, efforts are developing across regions and countries. In South Korea, Smart Construction program of government is an effort to join BIM-GIS systems on the too-centralized system applied both at permitting as well as infrastructure planning. For designs, automated code-checking engines guarantee that the design is ready according to national standards and scenario modeling tools let regional authorities evaluate the social-economic effects of large-scale projects before construction (Cantarero-García et al. 2023).

Through these cases, there are a number of best practices brought out. It was first noted that effective implementation is not merely related to technical integration, but also involves robust institutional coordination and delineated governance mechanisms of data sharing. Second, open data policy and easy-to-use visualization tools are the keys to developing the trust of people and participatory social decision-making. Third, the use of scenario simulation tools, especially related to climate resilience and infrastructure capacity, to augment BIM-GIS use and value to go beyond compliance and become proactive and adaptive planning.

These testimonies reveal that BIM-GIS integration is not a type of technology that fits all jurisdictions, and every single one will have to customize technology to its own legal frameworks, urban environment, and institutional culture. Yet, they also demonstrate that such systems can be employed in a considered way to increase greatly the accuracy, clarity and inclusivity of zoning and permitting activities and, ultimately, the way the next generation of spatial governance will arguably take place.

The final substantive section will look forward, outlining the future directions and innovations that could further expand the role of BIM-GIS in regulatory transformation.

8. FUTURE DIRECTIONS IN BIM-GIS FOR REGULATORY INNOVATION

BIM and GIS use in regulatory and zoning processes are still very much in the evolutionary stages with the most applications to date being limited to visualization, automatic compliance checks, and initial-stage scenario testing. However, the technological trend is shifting toward a more disruptive future that would see regulations themselves becoming dynamic data-driven agents that can adapt to changing conditions in the city. Some of these promising developments are indicative of the future in this direction (Mushtaha et al. 2024).

Another exciting angle is the combination between the BIM-GIS systems and artificial intelligence and machine learning to choose predictive zoning. Instead of responding to applications on an established regulatory basis, AI integrated models would be able to review past development trends, demographic patterns, climate records and infrastructure capabilities to suggest changes to zoning bylaws before issues occur. An example is that predictive models can be used to predict which neighborhoods will be congested or facing infrastructure pressure and interventions could take the form of regulation that is anticipated years before the issue occurs (Sarigul & Gunaydin, 2025).

Another innovation opportunity is through the spread of the Internet of Things (IoT). Real-time data can be fed into BIM-GIS systems via sensors installed in the building, infrastructure and environment and therefore constantly monitor energy consumption, air quality, noise level, and structural health. Such real-time data might go to adaptive regulations, in which environmental performance or occupancy limits adjust themselves according to the prevailing circumstances. These systems would transform governance based on passive code enforcement to the dynamic performance-based regulation (Nair et al. 2025).

The current environment of allowing and regulatory work has the prospect of increased transparency and accountability through blockchain technology as well. Records of regulatory checks, approvals, and amendments would be immutable and distributed in ledgers to which all participants have access that will diminish avenues of corruption, conflict, or data editing. Together with BIM-GIS automation, blockchain has the potential to facilitate a truly digital, end-to-end permitting environment in which all stages are verifiable and tamper-proof (Akindele et al. 2025).

Our increased exposure to immersive technologies like augmented reality (AR) and virtual reality (VR) is probably going to produce an increasing interest in regulatory review and engagement with the populace. AR-based inspection processes may potentially superimpose zoning boundaries and compliance cues onto the physical landscape, or VR models may enable policymakers and citizens to engage in virtual walkthroughs of proposed buildings and regulatory frameworks prior to policymaking (Azari et al. 2025).

The institutional aspect of innovation in the future is also vital. With further sophistication of BIM-GIS systems, the requirement to have cross-agency collaboration and similar data governance structures will become heightened. Imperative interoperability will compel more than technical compatibility of regulatory systems; it will necessitate plural municipal, regional, and national policies. Moreover, equitable access to these more advanced systems will have to be emphasized in order to prevent further widening of digital divides in planning and governance (Demir & Yomralioglu, 2024).

The history of the BIM-GIS innovation is heading towards a regulatory paradigm that is proactive, dynamic, and closely embedded within the lived experiences of an urban space. Adding predictive analytics, real-time monitoring, audit trails that cannot be altered, and immersive engagement tools, the regulatory frameworks of the future would transform into living things--armed with the rhythm of the city telling them how to change to suit their varying requirements (Lai et al. 2025).

The task that lies ahead will be to ensure innovation in these technologies does not undermine the practices of legality, fairness and democratic accountability that govern urban governance. In the event that such a balance is achieved, the forthcoming generation in BIM-GIS-enabled regulation no longer simply enforces rules but verses the principles of resilient, equitable and sustainable life in the city.

9. CONCLUSION

Zoning and regulatory systems are not new given they have been used in developing physical, social and environmental nature of cities. Traditional methods, however--which are based on written maps, written ordinances, and manual inspection--have found it difficult to keep up with the complexity, rapidity and interdependence of modern urban development. It is precisely the combination of Building Information Modeling and Geographic Information Systems known to address the challenge decisively since it provides the granularity of building-level information and the context of space intelligence.

Throughout the chapter, we have seen that BIM-GIS integration boosts the regulatory processes on several levels. Visual representation is the ability to transform the abstract zoning codes into spatial-rich forms of representation that are understandable even to the non technical audiences. Automated compliance systems add speed, consistency and predictability to the permitting effort, lessening administrative burdens and enabling a compliance-by-design approach. These capabilities have been further extended through digital twins into simulating policy, enabling regulators to experiment and predict consequences as well as track real-world compliance in real time.

Evidence gleaned in global case studies has shown that these technologies are not theoretical and purely a possibility because they are already in motion and already transforming governance in cities like Singapore, Helsinki, and Rotterdam, among others. These cases demonstrate just how important governance structures, institutional alignment and ethical protection not only go hand-in-hand with technical ability when it comes to adoption success. The near future is trending to a increasingly flexible regulatory paradigm enabled by artificial intelligence, live IoT data, blockchain secured audit logs and immersive community participation platforms.

Finally, the integration of BIM and GIS re-established policy, technology, and the built environment and transformed it in a way that is equivalent to a modernization of regulatory working processes. Cities are finally able to transition out of reactive law enforcement to proactive, participatory, and evidence-based governance by incorporating rules, context, and feedback loops into a single digital infrastructure. When applied transparently, inclusively, and legally-well grounded, such systems can transform not only into a tool to achieve compliance, but also one that can lead to stronger, fairer, and more sustainable cities in the future.

REFERENCES

Ahmad, M. (2023). AI-enabled spatial intelligence: Revolutionizing data management and decision making in geographic information systems. In *AI and Its Convergence With Communication Technologies* (pp. 137–166). IGI Global.

Akindele, O., Ajayi, S., Oyegoke, A. S., Alaka, H. A., & Omotayo, T. (2025). Application of Geographic Information System (GIS) in construction: A systematic review. *Smart and Sustainable Built Environment*, *14*(1), 210–236. DOI: 10.1108/SASBE-01-2023-0016

Al-Mhdawi, M. K. S., Dacre, N., Brito, M., Baxter, D., Xu, K., & Young, C. (2023, March). *An agile compliance framework for the European cooperation for space standardization. In 2023 IEEE Aerospace Conference*. IEEE.

Azari, P., Li, S., Shaker, A., & Sattar, S. (2025). Georeferencing Building Information Models for BIM/GIS Integration: A Review of Methods and Tools. *ISPRS International Journal of Geo-Information*, *14*(5), 180. DOI: 10.3390/ijgi14050180

Cantarero-García, G., Mollá, M., & González-Lezcano, R. A. (2023). Smart citizen in architecture and landscape. Method design based on spatial intelligence and universal accessibility learning by students with intellectual disabilities. *Frontiers in Built Environment*, *9*, 1094760. DOI: 10.3389/fbuil.2023.1094760

Cao, Y., Xu, C., Aziz, N. M., & Kamaruzzaman, S. N. (2023). BIM–GIS integrated utilization in urban disaster management: The contributions, challenges, and future directions. *Remote Sensing (Basel)*, *15*(5), 1331. DOI: 10.3390/rs15051331

Castellanos-Ardila, J. P., Gallina, B., & Governatori, G. (2021). Compliance-aware engineering process plans: The case of space software engineering processes. *Artificial Intelligence and Law*, *29*(4), 587–627. DOI: 10.1007/s10506-021-09285-5

Chiodelli, F., & Morpurgo, D. (2022). Placing the law: The socio-spatial impact of legal norms beyond mere compliance. *Transactions of the Institute of British Geographers*, *47*(3), 725–740. DOI: 10.1111/tran.12530

Chong, H. Y., Ma, Q., Lai, J., & Liao, X. (2025). Achieving Sustainable Construction Safety Management: The Shift from Compliance to Intelligence via BIM–AI Convergence. *Sustainability (Basel)*, *17*(10), 4454. DOI: 10.3390/su17104454

de Vries, W. T. (2023). Ethical and Legal Issues Affecting Geospatial Sciences and Advances in Smart Land Management. In *Geospatial Science for Smart Land Management* (pp. 421–430). CRC Press. DOI: 10.1201/9781003349518-31 del Cerro Velázquez, F., & Morales Méndez, G. (2021). Systematic review of the development of spatial intelligence through augmented reality in stem knowledge areas. *Mathematics*, *9*(23), 3067. DOI: 10.3390/math9233067

Demir, S., & Yomralioglu, T. (2024). Bridging geo-data and natural gas pipeline design standards: A systematic review of BIM-GIS integration for natural gas pipeline asset management. *Energies*, *17*(10), 2306. DOI: 10.3390/en17102306

Gonthier, C. (2022). Cross-cultural differences in visuo-spatial processing and the culture-fairness of visuo-spatial intelligence tests: An integrative review and a model for matrices tasks. *Cognitive Research: Principles and Implications*, *7*(1), 11. DOI: 10.1186/s41235-021-00350-w PMID: 35119577

Hu, J., Lv, Z., Yuan, D., He, B., Chen, W., Ye, X., Li, D., & Yang, G. (2023). A spatiotemporal intelligent framework and experimental platform for urban digital twins. *Virtual Reality & Intelligent Hardware*, *5*(3), 213–231. DOI: 10.1016/j.vrih.2022.08.018

Lai, J., Wan, R., Chong, H. Y., & Liao, X. (2025). Digital Intelligence in Building Lifecycle Management: A Mixed-Methods Approach. *Sustainability (Basel)*, *17*(11), 5121. DOI: 10.3390/su17115121

Li, D., Shao, Z., & Zhang, R. (2020). Advances of geo-spatial intelligence at LIESMARS. *Geo-Spatial Information Science*, *23*(1), 40–51. DOI: 10.1080/10095020.2020.1718001

Lindgren, D. (2020). Framework for assessing compliance. In *An Assessment Framework for Compliance with International Space Law and Norms: Promoting Equitable Access and Use of Space for Emerging Actors* (pp. 49–84). Springer International Publishing. DOI: 10.1007/978-3-030-15762-3_5

Liu, Y., Liu, Y., Chen, X., Lin, C., & Zhou, Z. (2025). Balancing spatial logic and governance objectives for street vending: An optimizing model for dynamic formalization zones. *Cities (London, England)*, *160*, 105819. DOI: 10.1016/j.cities.2025.105819

Lorenzo-Eiroa, P. (2022). Space-environment commons: From big data survey to AI, to a post-capitalist blockchain zoning platform. In *Resilient Communities and the Peccioli Charter: Towards the Possibility of an Italian Charter for Resilient Communities* (pp. 161–172). Springer International Publishing. DOI: 10.1007/978-3-030-85847-6_17

Martin, A. S., & Freeland, S. (2021). The advent of artificial intelligence in space activities: New legal challenges. *Space Policy*, *55*, 101408. DOI: 10.1016/j.spacepol.2020.101408

Martino, N., Sanders, C. B., & Dej, E. (2024). The 'regulatory grey zone': Bylaw enforcement's governing of homelessness and space. *Policing and Society*, *34*(4), 305–320. DOI: 10.1080/10439463.2023.2263617

Mushtaha, A. W., Alaloul, W. S., Musarat, M. A., Baarimah, A. O., Rabah, F. K., & Alawag, A. M. (2024). BIM-GIS Integration for Infrastructure Management in Post-Disaster Stage. In *2024 ASU International Conference in Emerging Technologies for Sustainability and Intelligent Systems (ICETSIS)* (pp. 856-861). IEEE. DOI: 10.1109/ICETSIS61505.2024.10459527

Nair, S. M., Andrushia, D., Carmichael, M. J., Ebenezer, J. S., & Varghese, A. (2025). GIS and BIM integration for assessing and maintaining existing buildings: a review. *Proceedings of the Institution of Civil Engineers-Forensic Engineering*.

Nica, E., Popescu, G. H., Poliak, M., Kliestik, T., & Sabie, O. M. (2023). Digital twin simulation tools, spatial cognition algorithms, and multi-sensor fusion technology in sustainable urban governance networks. *Mathematics*, *11*(9), 1981. DOI: 10.3390/math11091981

Pelden, S., Banihashemi, S., Mohandes, S. R., Arashpour, M., & Kalantari, M. (2025). Enhancing infrastructure planning and design through BIM-GIS integration. *Structure and Infrastructure Engineering*, •••, 1–20. DOI: 10.1080/15732479.2025.2461629

Razinkina, I., Bulatenko, M., Chernov, S., & Prasolov, V. (2022). Ethical and legal balance of modern economic intelligence. *The Journal of World Intellectual Property*, *25*(2), 335–346. DOI: 10.1111/jwip.12225

Sarigul, F. H., & Gunaydin, H. M. (2025). *Integrated BIM, GIS and interoperable digital technologies in lifecycle management of building construction projects: systematic literature review*. Smart and Sustainable Built Environment.

Sullivan, R. (2022). The Geopolitics of Urban Digital Twins: Spatial Cognition and Predictive Modeling Algorithms, Simulation and Virtualization Technologies, and Blockchain-enabled Cyber-Physical Systems. *Geopolitics, History, and International Relations*, *14*(2), 25–39. DOI: 10.22381/GHIR14220222

Tan, Y., Liang, Y., & Zhu, J. (2023). CityGML in the Integration of BIM and the GIS: Challenges and Opportunities. *Buildings (Basel, Switzerland)*, *13*(7), 1758. DOI: 10.3390/buildings13071758

Vergara-Perucich, J. F. (2025). AI-Driven Deconstruction of Urban Regulatory Frameworks: Unveiling Social Sustainability Gaps in Santiago's Communal Zoning. *Urban Science (Basel, Switzerland)*, *9*(6), 186. DOI: 10.3390/urbansci9060186

Wang, J., Mo, F., Qiao, S., Feng, H., & Lyu, Z. (2025). Spatial computing in digital twins. *Digital Twin*, 2508268.

Wang, K. (2021). A theoretical analysis method of spatial analytic geometry and mathematics under digital twins. *Advances in Civil Engineering*, *2021*(1), 8910274. DOI: 10.1155/2021/8910274

Wang, X., & Xie, M. (2024). Integration of 3DGIS and BIM and its application in visual detection of concealed facilities. *Geo-Spatial Information Science*, *27*(1), 132–141. DOI: 10.1080/10095020.2022.2054732

Wang, X., Yang, J., Han, J., Wang, W., & Wang, F. Y. (2022). Metaverses and DeMetaverses: From digital twins in CPS to parallel intelligence in CPSS. *IEEE Intelligent Systems*, *37*(4), 97–102. DOI: 10.1109/MIS.2022.3196592

Wu, D., Zheng, P., Zhao, Q., Zhang, S., Qi, J., Hu, J., & Wang, L. (2026). Empowering natural human–robot collaboration through multimodal language models and spatial intelligence: Pathways and perspectives. *Robotics and Computer-integrated Manufacturing*, *97*, 103064. DOI: 10.1016/j.rcim.2025.103064

Xie, Y., Liu, J., Hu, N., Li, D., Song, Z., & Zhang, Z. (2025). Towards Ethical Spatial Decision-Making in Geospatial Artificial Intelligence. In *GeoAI and Human Geography: The Dawn of a New Spatial Intelligence Era* (pp. 313–325). Springer Nature Switzerland. DOI: 10.1007/978-3-031-87421-5_22

Yau, S. S., Pandya, K., & Choudhary, S. (2024, July). Regulatory Compliance in Software Services Using Emerging Technologies. In *2024 IEEE International Conference on Software Services Engineering (SSE)* (pp. 36-42). IEEE. DOI: 10.1109/SSE62657.2024.00018

KEY TERMS AND DEFINITIONS

Building Information Modeling (BIM): A digital representation of the physical and functional characteristics of a building or infrastructure, containing detailed 3D geometry, material specifications, and operational data in an object-oriented, information-rich model.

Compliance Automation: The process of using computational rule engines and AI-powered systems to automatically verify whether a proposed design meets applicable regulations and building codes.

Data Governance: The set of policies, procedures, and standards that define how data is collected, stored, shared, and protected to ensure accuracy, privacy, and ethical use.

Digital Twin: A dynamic, digital replica of a physical asset, building, or city that integrates real-time data to simulate, monitor, and analyze performance and environmental interactions.

Geographic Information Systems (GIS): A spatial data framework used to capture, store, analyze, and visualize georeferenced information about the earth's surface and human-made environments.

Interoperability: The capacity of different digital systems, platforms, or organizations to exchange and use information seamlessly, without loss of accuracy or meaning.

Permit Processing: The official review and approval workflow required before a construction or land-use project can proceed, often involving multiple agencies and compliance checks.

Policy Scenario Simulation: The use of virtual modeling tools to test potential changes in regulations or urban plans, assessing their likely spatial, environmental, and socio-economic impacts before implementation.

Predictive Zoning: The application of data analytics and machine learning to forecast future land-use needs and proactively adjust zoning regulations in anticipation of demographic, economic, or environmental changes.

Regulatory Framework: The collection of laws, codes, ordinances, and policies that govern land use, building construction, environmental compliance, and urban development.

Smart City: An urban area that uses digital technologies and data-driven systems to improve the efficiency of services, infrastructure, governance, and quality of life for residents.

Spatial Decision Support System (SDSS): A computer-based platform that integrates spatial data and analytical models to assist in decision-making for urban planning, infrastructure investment, and policy design.

Spatial Intelligence: The ability to derive meaningful insights and make decisions by integrating and analyzing spatial, environmental, and socio-economic data.

Zoning: A legal tool used by governments to regulate land use and development intensity, typically by designating specific zones for residential, commercial, industrial, or mixed uses and setting parameters such as height limits, setbacks, and floor-area ratios.

Chapter 7

Sustainable Urban Design and Management: The Role of BIM and GIS Integration

Jyoti Rani
https://orcid.org/0000-0002-5577-7808
Parul University, India

Shashiranjan Kumar Yadav
Lovely Professional University, India

ABSTRACT

Sustainable urban design and management are essential to address rapid urbanization, climate change, and resource depletion. This chapter examines the integration of Building Information Modeling (BIM) and Geographic Information Systems (GIS) as a comprehensive approach to creating smarter, greener cities. BIM offers building-level precision, while GIS provides spatial analytics and large-scale environmental context. Their integration enables informed decision-making, efficient resource use, and resilient infrastructure development. The chapter outlines theoretical frameworks, advances in 3D modeling and data interoperability, and real-world applications in green infrastructure, environmental impact assessment, and sustainable land use planning. Emphasizing BIM–GIS's role in advancing sustainable design principles, it serves as a guide for researchers, practitioners, and policymakers to adopt integrated spatial intelligence tools for sustainable urban transformation.

DOI: 10.4018/979-8-3373-4058-6.ch007

INTRODUCTION

Urban design and development, the sustainability evaluation of which is considered essential for directed decision-making, A sustainability assessment tool also acts as a catalyst in promoting sustainability practices by acknowledging good performers through their rating system and this realization creates a market pull towards sustainable products and processes (Ameen et al., 2015). Urbanization also causes significant shifts in the use of natural resources. It would facilitate the transition from conventional to modern fuels but it is rather intensifying the use of energy and its environmental impacts (Alberti & Susskind, 1996). According to (Keivani, 2009) the next 20 years in cities of the developing world will absorb over 95% of urban growth. Furthermore, fast urbanisation has blossomed into the emergence of developing country-based megacities of over 10 million. In 1975 we had three megacities: Tokyo, New York and Mexico City. By 2005, there were 20 such cities with at least one million inhabitants and all but four were in the developing world (WUP)[REMOVED HYPERLINK FIELD].

While Fay and Opal had debated that urbanisation rate is not directly associated with economic growth in its cities, but it inversely related to the level of urbanisation in a nation especially in the African context. This describes perfectly the process as christened by them: Urbanisation without Growth, from an analysis of 187 cities in 100 countries. sustainability in urban design of a city cannot be realized only through the evaluation of isolated buildings or even building synthesis, but it needs to take a comprehensive approach to the whole city and all its miscellaneous parts (like neighbourhoods with their components, population mobility transportation land use urban spaces water usage energy usage geographical diversity air quality etc,) as well. These are the essential components that comprise the bedrock of urban sustainability assessment.

Indeed, rapid urbanization is anticipated to lead many countries to see their share of the urban population increase between now and 2025; India and China both are expected to have about one-third more urban population in a decade. Based on the three trends number urban build-up is spint with fluctuation economies of scale together with diseconomies like leakages as well as environmental degradation Cities are faced with balancing between the pros and cons of urbanization from environmental, social and economic aspects (Verma & Raghubanshi, 2018). Urbanization causes deteriorated air and water quality, consumes more resources and stresses natural environment, energy resources as well as infrastructure.

Addressing these challenges demands a consideration not only for the city a nexus of environmental context, built infrastructure, and human communities but also calls for collaboration among scholars and practitioners across various disciplines. Cities all over the world are more and more connected under the trend of globalization, making each path they take and experience in handling urban challenges precious one to another (Li et al., 2018). Unlike the situation in Western Europe and North America where acelerated urbanization set in, one outstanding feature of this rapid "Third World" urbanization is that it is most noticeable at the lower not highest level of economic development which means industrialisation lags far behind the rate of urbanization (Ilesanmi, 2010).

Urban planning is inherently connected to the history, current state and futures of all urban communities, making it a subject of immense interest to researchers and practitoners. Sound urban planning is inevitably intertwined with the economy, welfare, and their orientation to the right development horizon and sustainable community-based growth (Elsheikh et al., 2021). Over 40% of the urban population of the world lives in towns and cities with populations fewer than 500,000; approximately one eighth live in megacities (over 10 million inhabitants). Among the places that include remarkable urban centers such as Tokyo, Delhi, Shanghai, Cairo and Mumbai where each city has a very minimum of 20 million residents expresses how massive their surroundings influences the economic surrounding, ecological conditions and social aspects (Marzouk & Othman, 2020).

Importance of Integrating BIM and GIS for Sustainable Urban Development

The subject of energy efficient buildings incorporated into the district is one of the greatest and most challenging of research priorities in the European Union. In this context, the potentiality and benefits of the integration of Geographic Information System (GIS) – Building Information Modelling (BIM) are explored as an integrated methodology for data collection and management relating to existing buildings. GIS technology has provided a new environment for planning, including the possibility of linking digital and visual data, and stratification of a large group of important information that is adopted when planning, such as infrastructure services, road paths, ownership boundaries, river paths, avoiding heritage, sharp trends and other planning determinants that it became easy to deal with GIS. GIS supports the important spatial analysis theory where Spatial analysis allows you to solve complex problems related to location, find patterns, evaluate trends, and make decisions(Elsheikh et al., 2021).

Figure 1. GIS and Urban planning

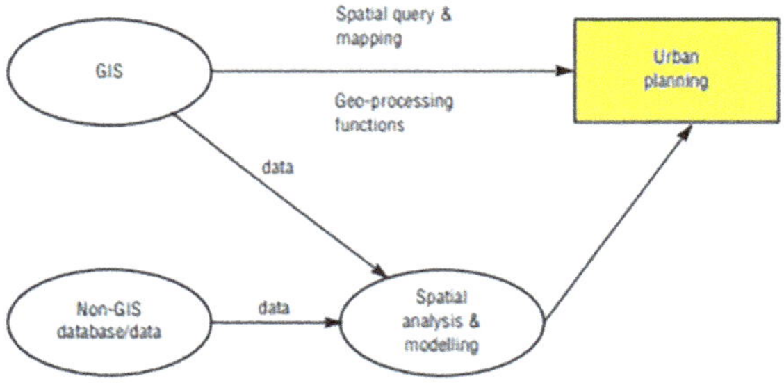

In the developed and as well in developing countries GIS has been used widely in urban planning. Planners use GIS tools to analyze problems quicker and more comprehensively, develop solutions, and track themselves on a path towards the long-term goals of building communities (Paul, 2018).

BIM (Building Information Modelling) can also be not a tool for urban but powerful as the nature of it as this has an ability that to combine data with a physical form and enables us to joins planning regulations up together with the city. For the public and private sector, BIM visualizations function as a communication tool, helping them to understand complex urban design challenges.

It would not be possible to model a large-scale urban area with the existing tools in BIM methodology. GIS and BIM integration has ushered in a new chapter of various applications in cityplanning: such as the construction, transportation through roads, infrastructure services, space utilization modeling, future planning model... urban development activities, engineering informatics, waste management Cairo garbage collection and disposal methodologies; Water supply for water supplies spatial analysis; Management (Management of archaeological sites), etc.

The model used was by merging the BIM and GIS in building operate management including all different kind of facilities. They include the model for raising the level of efficiency in space management, discovering collision of management and how to manage it, complete transportation system modeling with its different systems, reality simulation and etc.

visioning the future, researching lighting and exploring ambiguous questions related to the intended project as well as its potentials opportunities. Using Building Information Modeling (BIM) and Geographical Information System (GIS) to design and predict the utility infrastructure needs of the expanding and developing cities, which reveals smartness from planning stage. The infrastructure needs that we consider are: Fresh water withdrawal, sewage capacity and electricity use. The city is, on the one hand, divided into several districts and lands and plots (Marzouk & Othman, 2020).

By adding the potential of both BIM and GIS technologies, a single tool to manage object during full life cycle in different contexts and specific territory. Implementation of these systems within the integration framework can reduce time required for decisions of management level and provide an in-depth analysis at the pre-investment stage during development of a construction project (Sheina et al., 2023). Over recent years, a significant change has taken place within the field of ICT for the construction industry in which Building Information Modelling [BIM] is now one of the most active topics in both industrial and academic spheres as an alternative Computer Aided Design (CAD) paradigm (Bryde et al., 2013).

The statement also indicates that GIS and BIM data must be considered decisively for smart and sustainable cities during the entire life cycle, as mentioned above. The BIM model records accurate, updated and detailed graphical as well descriptive data of all elements involved in a construction. This data is paramount to everyone from the designer, the owner, the ultimate consumer and the person who serves as decision support. The data in GIS systems of which all the instances describe the same building from different perspectives is unambiguously connected to its actual location on the city or region map. For instance, a building site can be observed in the view of buildings in its environment, its areal extension can be measured accurately with distances for services positioned and integrated into systems as GIS to allow better visualization beige information from satellite imagery. Everywhere, reliable and up-to-date spatial information is crucial throughout the entire scope of environmental management and sustainable development.

Figure 2. Critical review of BIM or GIS

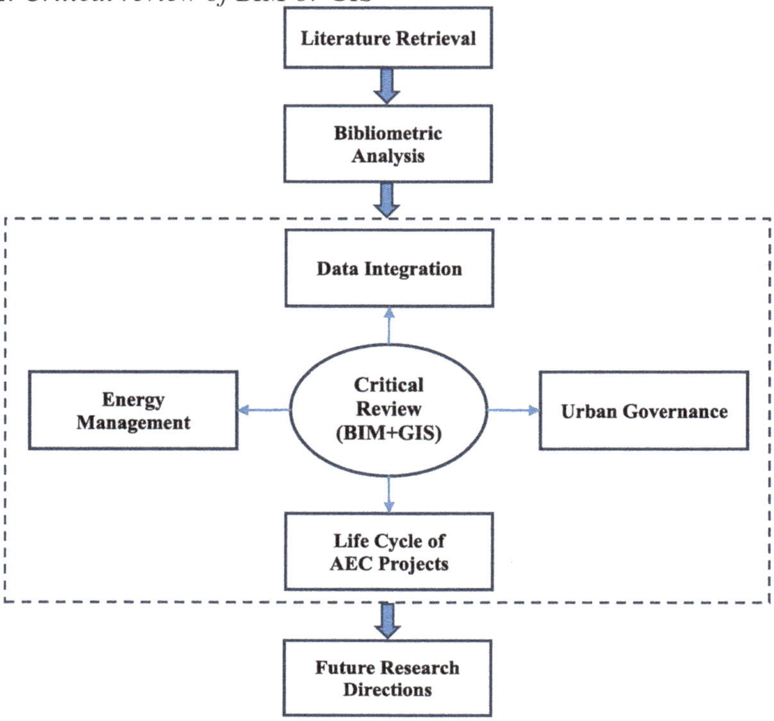

Based on different dominant positions, three typical modes of BIM-GIS architecture can be generalized: BIM leads and GIS supports, GIS leads and BIM supports, and BIM and GIS are equally involved. Moreover, the areas of ongoing research and future work are suggested, paving the way to enhance BIM-GIS integration for their wide-ranging applications in the sphere of sustainable built environment (H. Wang et al., 2019). Optimal Technical and Policy Solution for Realigning City Infrastructure GIS-BIM based energy planning system Section 2 Firstly, it depicts the Japanese model of smart city from urban planning to multiple infrastructures (Yamamura et al., 2017). This makes it possible to carry out complex analysis including energy simulations, traffic flow and environmental impact studies. It has also helped in enhancing the working of planners, developers and other stakeholders so that a better urban development strategy can be prepared (ElBeheiry, 2025).

The integration of the GIS and BIM data platforms can build a technological foundation for cross-field data integration, break through the field barriers in engineering stages in the process of urban planning, smart city redevelopment, construction and renovation, ACE industry, urban rail transit industry and municipal Engineering to address large-scale problems Waste of value data (Xia et al., 2022).

sustainable (green) performance driven economic benefits. buildings that Stop the Environment According to (Jalaei, 2015) changing parameters at the green design such as cost of components reused in building, life cycle assessment and energy consumption, characteristics of light and impact points of green rating during conceptual design help designers for decision-making with regard to less environmental degradation due to other alternatives. Building Information Modelling (BIM), is the modern approach and combine it with green and smart planning objects of buildings, neighbourhoods, cities (Dall'O' et al., 2020). BIM is the more granular buildings themselves purpose and GIS is more of larger picture on what surroundings are available (macro level). Similarly, a holistic view of the built environment, integrated with all the additional required data is one such solution that educates like architecture, engineering and construction (AEC) industries for an informed path in building integration and transitioning to digital age.

BACKGROUND AND CONTEXT

Smart city is a vital innovative concept and urban development mode that must be applied in urban scientific management and development to realize sustainable. These are a tangible expression of profound integration between new generation of information technology applications and urban economic and social development. The construction or reconstruction of smart cities has led to sustainable urban development (Xia et al., 2022).

Historical Development of BIM and GIS Technologies

Data integration, application in the life cycle of AEC projects, building energy management, Urban governance technologies. Therefore, (H. Wang et al., 2019) position three common modes of BIM-GIS integration, specifically "BIM leads while GIS supports", "GIS leads while BIM supports", and "equally involved for the two sides" based on the main role carried in both technologies.

There are three main aspects that must be considered more pertinent: (i) the first is related to an intelligent data modeling of GIS and BIM into one environment; (ii) together with the ability to include historical period data for time-based analysis (4D); and, finally, integrating structured datasets from pre-existing HGIS or HBIM systems (Bruno et al., 2020).

This blends old, new historical,Architectural Survey and building material data to makeGIS Based BIM Platform. We will achieve that using data integrated within 2D GIS thematic maps complemented with a development of 3D HBIM (Tsilimantou et al., 2020). A BIM is a library of digital information in open and protected formats which captures graphical information together with nongraphic data linked to that graphical element. There are two faces: to BIM created together with the project, and for the project designed directly in BIM (as it is gear of new objects designed in CAD systems for example, Revit software), or old/historical buildings. The second is a technological system that has been modern and globally used today. The problem from the point of view of the engineer is the creation of BIM for legacy objects (Pavelka et al., 2019).

Provide stakeholders with the automated and digital information necessary to carry out a wide range of tasks, directly linked to the building's 3D-BIM modeltranslator (support the language): The developed system (applicability and functionality tests) was tested in a building of significant heritage value — the Casa de Santo António, located in Ílhavo, Aveiro, at the centre of mainland Portugal. Microsoft SQL Server® was used to create and maintain the database of the management system, which would accommodate all data exported from BIM model (Rodrigues et al., 2019).

Figure 3. Historical building information modelling flowchart

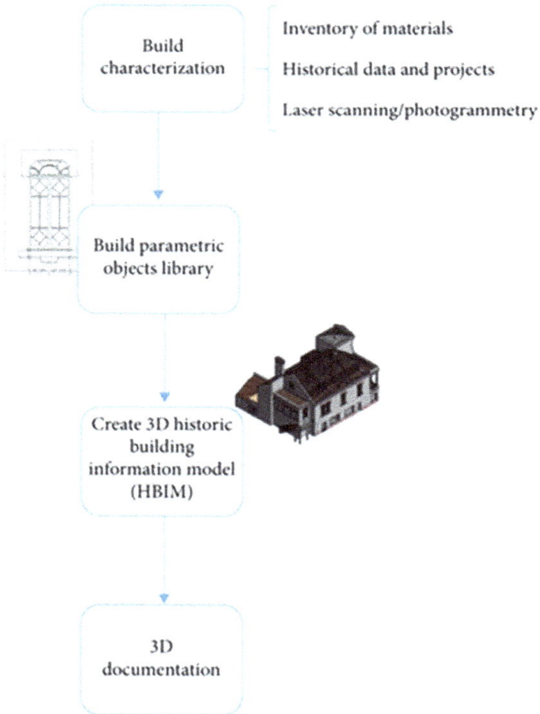

Today, GIS and BIM are utilized in different areas: the former being more common for generally monitoring the structures in city and their land uses or risk analysis for environment or hazards while latter is applied on planning new building and infrastructures (Saccucci & Pelliccio, 2018).

Figure 4. Bibliometric Analysis of Literature relevant to BIM, Landscape, Planning, Urban Design and Climate Change (Jain et al., 2022).

Figure 5. BIM-GIS Platform for Infrastructures

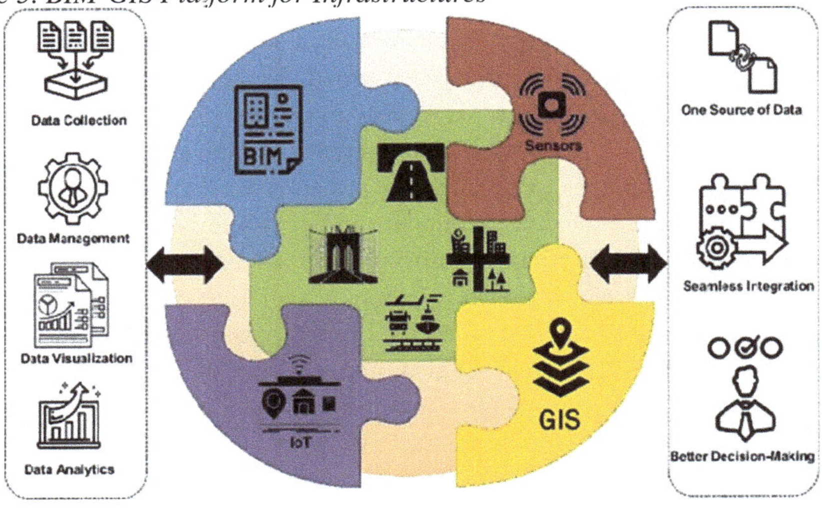

THEORETICAL FRAMEWORK FOR BIM AND GIS INTEGRATION

Principles of Integration for Sustainable Urban Planning

Urban design is between the layers that allow this integration to be carried out, constructing everything Medieval infrastructure for cultural diversity and needs. With urban creativity, Hard and soft infrastructure are the necessary pre-conditions, according to Healey "hard" means rules – procedures– protocols, "solid" networking of social norms based on trust and reciprocity. Urban design is viewed at a scale of production as a process of place creation differs with these multiple perspectives hence combining differences different dimension regarding the space production (He et al., 2020).

Principles of integrated approach to various: social, architectural, artistic, economic, sanitary and hygienic demands. Societal wants are to make the connection and sewerage pipeline system networked from which individuals take maximum benefit in the construction of their buildings. The architectural-artistic side of the solutions focuses on single spatial composition with surrounding landscape.

Economic requirements are such as to require reasonable use of area territories. Sanitary hygiene demands are of higher interest when it comes to sustainable urban areas: provision of an effective insulation and air conditions in all area, as well as protection of buildings from the noise coming from outdoors (Alhazmi & Loy, 2021).

Footnote: development plan in southern climatic zone (taking into account influence of residency mass), locality and condition, constraints on the place of future construction This sustainable development allows for the fabrication of new architectural-engineering principles and methods. The architecture of a building that is economical in terms of energy falls under this. In relation to the analysis of experience in design of energy efficient residential buildings both foreign and domestic, a list of requirements can be established (Leyzerova et al., 2016).

Urban planning and design can play a very important in creating such sustainable growth by including ‗sustainable development principles' in the form of urban planning strategies, policies, programs and projects. Sustainability has different conceptualization, interpretation and understanding; there is no single format of incorporation sustainability principles in a holistic urban planning (Al-Shihri, 2013). Urban planning adapts continuously to complex social processes and requires a methodology supporting multidisciplinarity, flexibility and adaptability (Milojević & Kuvač, 2023). This study is a novel quantitative review of key sustainable urban neighborhood development principles from the sustainability and urban literature. We, therefore, (i) recognize the promising pathways with principles that are near to the core sustainability criteria and (ii) seed foundations for more comprehensive set of principles by compiling this reviewed literature. And, lastly (iii), the lacking consideration of sustainability aspects is pointed out. Based on the findings (iv) a road map is drawn for advancing the theoretical development of a sufficient set of fundamental sustainability principles for urban neighborhood development (Luederitz et al., 2013).

Figure 6. Rational urban planning process (Yigitcanlar & Teriman, 2015)

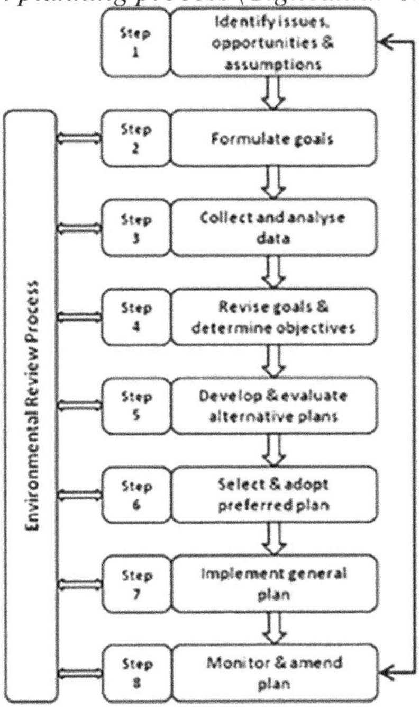

he rational, urban planning approach in the a priori stage ensures clear systemic and structural relation between each step of process; logic and evidence using for analising planning issue as well as providing more reasonable way to protest the future by its continue review along each steps. Uurban planning processes are well-suited to provide systematic direction on the interrelations between human activities and ecosystem health. Emphasizing the need to consider the context of ecology in policy and decision-making or evaluating societal interaction amongst human–natural (Qiu et al., 2021).

Figure 7. Interdependent components of ecosystem approach

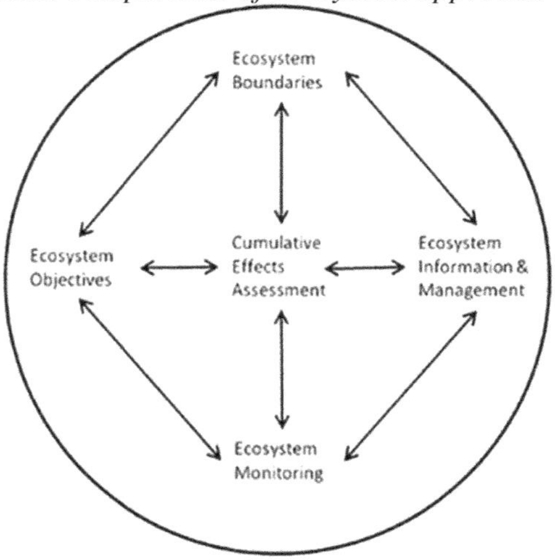

Figure 8. Event-based development process

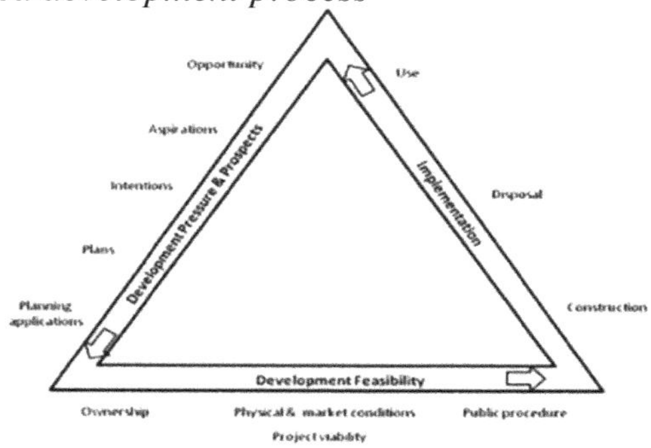

Figure 9. Proposed integrated urban planning and development process

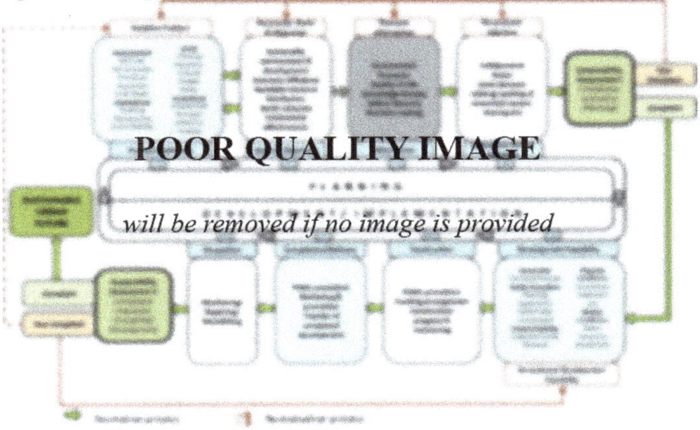

A fresh perspective on rethinking SUD includes an integrated urban assessment, planning, development and management process that is environmentally integrated using sustainability approaches (J.-M. Liu et al., 2023). In the present paper, by refering to sustainability principles in Curitiba Iran it tries to investigate their applicability on Shiraz as a case study. Conclusion; In view of the inherent characteristics and systems in urban planning, both physical, social, cultural and environmental; this research has leant huge success for contextual studies while sustainability is achieved through integration and implementation strategies at urban scale (Bueno et al., 2013).

- **Data interoperability and standards enabling BIM-GIS synergy**

The methods for attaining interoperability of BIM and GIS tools differ. Examples of such initiatives include adapting the tools or standardizing processes, utilizing open data standards(Graziano et al., 2023). Interoperability gaps that need to be bridged to allow design & construction information to connect with operational asset information. The Crossrail application of GIS and BIM systems was investigated in relation to some of the challenging problems that can appear during the course of one infrastructure asset life cycle (Elshaboury et al., 2025).

Spatial Intelligence as a Foundation for Sustainable Design Decisions

Interoperability of BIM and GIS tools is achieved through different methods. For example, by adjusting the tools or establishing standard processes to use open data standards (Kang & Hong, 2015). One of the primary advantages is that it allows you to view virtually any architectural or environmental entity in a digital realm, controlling information applicable in space, providing a transitional direction combining spatial data with different levels of detail and allowing for an effective standardized methodology for planning construction development. Analyzes space and organizes layers of information into map visualizations and 3D scenes. By providing this singular functionality, GIS exposes deeper insights within data that includes patterns, relationships and context; driving smarter decision making.

TECHNOLOGICAL FOUNDATIONS

Software Platforms and Tools Supporting BIM and GIS Integration

Software architecture for the effective integration of building information modeling (BIM) into a geographic information system (GIS)-based facilities management (FM) syste a software architecture to help the integration of building information model (BIM) into geographic information system (GIS)-based facilities management (FM) systembase on (Kang & Hong, 2015).

Figure 10. Example of building models designed in the SketchUp software.

The principal cause was the architects selected design software and updates to detailed architectural designs. Using the reduced footprint a 2D urban plan was created conforming with the reduction (offset) values, and on this offset based plan, 3D building models was put on manually (Buyukdemircioglu & Kocaman, 2020).

Advances in 3D City Modeling and Visualization for Sustainability

CSDILA Platform is a 3D visualization and modeling platform proposed in this work– it was assigned to model and visualize different dimensions so as to reach sustainability, specifically in urban context by (Amirebrahimi & Rajabifard, 2012).

Here, the mentioned GIS, SD model and 3D visualization integration is in reference to GISSD system where the interaction and variation of the sustainability indicators for residential development can be best explained The paper structure is as follows: Section 2 gives brief introduction about GISSD system and summarized a general application of System Dynamics (SD), GIS and 3D visualization (Xu & Coors, 2012). This paper uses a systematic review to articulate the challenges and utility of 3D models and visualization in urban development. In infrastructure and urban development domains, either 3D is being used or the use of a 3-D visualization is propagating due to which it became necessary for the industry to have an approach to utilize this high range results. 3D visualization is more used for design. And in 3D environments, artists and architects want to view high resolution detail crafting environment. The research about parametric design shows that 3D modeling of building information models (BIM) made in a generative and iterative way can contribute to bettering output detailed plans in architecture and urban planning.

Figure 11. 3D modelling (Buyukdemircioglu & Kocaman, 2020)

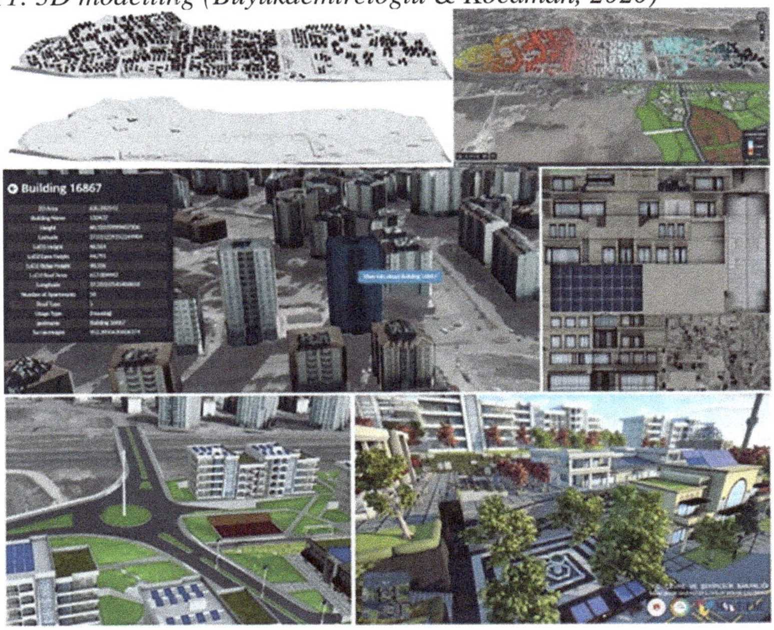

In turn, the 3D models of urban structures (buildings, city furniture, plans, vegetation or infrastructure, among others) are part of a 3D GIS (geographical information system), considered an indispensable technology for designing and developing high-value-added smart cities that essentially require site-specific design. The applications of urban simulation and visualization include regional planning, transportation, land-use regulations, environmental protection etc. But as we get to almost complete virtual 2D coverage of whole city areas/structures it gets increasingly harder to allow us a perception on more complex urban structures. With out abstract visualization, weare able to hide the complexity of these models where its visibility is not required, while leaving features that are significant draw emphasis for better understanding and communication (Glander & Döllner, 2009).

Figure 12. Examples of building visualizations: (1) Textured 3D building model; (2) 3D building footprints; (3) 2D Building footprints; and (4) Image from Google Street (Johansson et al., 2016).

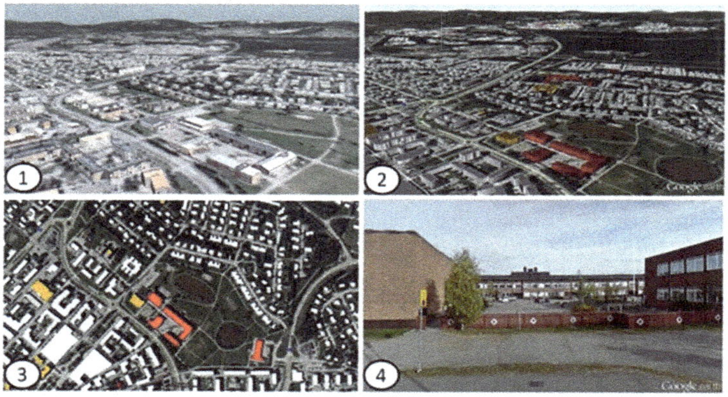

Compared with 2D-visualization, the cognitive understanding of vertical dimension and spatial relation increases when visualizing the 3D models (especially when you can walk through the whole 3D environment). In these contexts, they are more appropriate than 2D models because they help to gain qualitative understanding and can provide a more universal representation of buildings that support communication between stakeholders with different education levels and professional backgrounds.

Figure 13. 3D city models may be applied in a multitude of application domains for environmental simulations and decision support (Biljecki et al., 2015).

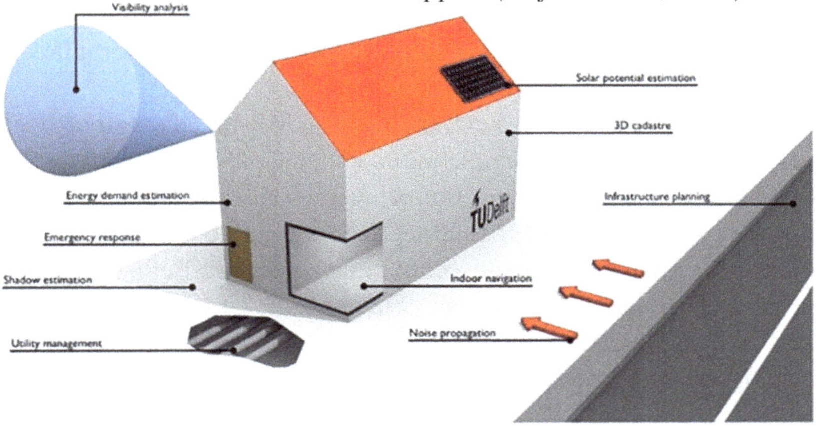

Acquired from different CityGML range acquisition techniques in a 3D city model, such as photogrammetry and laser scanning which are commonly known perception data, extrusion from footprints, synthetic aperture radar (SAR), architectural models & drawings, handheld devices simulation-based approaches for buildings procedural modelling and volunteered geoinformation approach which aiming to serve collaborative mapping through social web participate. Until now it appears that the early applications of 3D city models were dominated by visualisation. In urban and built-environment visualisation the adoption of 3D functionalities brings out this information in a complete picture which makes delivery intuitive. 3D urban models have a lot of parameters like the type of client, application output,his expected output,budget,time period and area to be covered (Shiode, 2000).

Techniques for Real-Time Data Sharing and Collaborative Workflows

From the definition of Kuan and Szweda you see that a workflow is an activity in which coordinated work on tasks by separate processing elements is done [KS 95]. This tasks could be manual, or automated, either developed explicitly for the workflow application we are developing, or might have been already existing as legacy programs A workflow is an automated organizational process integrating human (manual) and automatic task composites (Sheth & Kochut, 1998). Applications of VR have focused on design coordination, project planning, construction education and safety training, construction operations coordination (both on the field with workers and in control rooms with planners), facility management, or real estate

[26]. A popular workflow of AEC/FM VR modeling is to create the BIM model (such as a model from Revit) into a.fbx file or other graphing files by using third party graph programs (for example, 3D studio Max), and then put the processed data into game engines (such as Unity 3D), which will finally render it in VR mode while running programly (Du et al., 2018).

B I M models, and feature-based tracking in an MR-based distributed situation to let the workers know about the time-aware location of a field worker can retrieve information while visualising inspection & maintenance operations, attuning real-time responses besides supporting field service level collaboration enabling virtual communication among both office-executive managers & on-field service deliverers (El Ammari & Hammad, 2019). From 1992 onwards, an increasing literature has developed on the application of GIS to support collaboration among such groups, and even to foster more knowledgeable and active citizen participation in decision-making activities.

Being an integrated system to model amazing mapping and space data, GIS has always been a common spatial decision made because of the unique features they provide. But also hinders once a traditional GIS single- user interaction mode of the problem-solving complexity and poor efficiency. Although collaborative GIS is envisioned as workgroup computing, integration with communication (email, real-time chatting, discussion forum and conferencing) and process coordination (workflow) tools will be increasingly essential for a seamless collaboration rather than visiting the different types of systems to perform these basic tasks (Sun & Li, 2016).

APPLICATIONS IN SUSTAINABLE URBAN DESIGN

The significance of the Environmental Impact Assessment (EIA) on the design process will increasingly become a matter of key concern for both public and private sectors, as design decisions have profound long-term impacts upon environmental performance of infrastructure projects. LCAs, as outlined in the ISO 14040 series of standards, assess the environmental effects related to a product (of service or process) at every stage throughout its lifecycle. The American Society of Civil Engineers (ASCE) provides the following definition relating to sustainability, which is the only inclusive term in common usage: SUSTAINABILITY Sustainability is a set of environmental, economic, and social conditions in which all of society has the capacity and opportunity to maintain and improve its quality of life indefinitely without degrading the quantity, quality or the availability of natural economic and social resources. This was the beginning of the Civil Engineering Environmental Quality Assessment and Award Scheme (CEEQUAL) from the Institution of Civil

Engineers in England. Finally, a normalized average environmental impact range score of each item is determined and visualized in the BIM model with heat map style.

Figure 14. Proposed framework BIM-based environmental impact assessment for infrastructure.

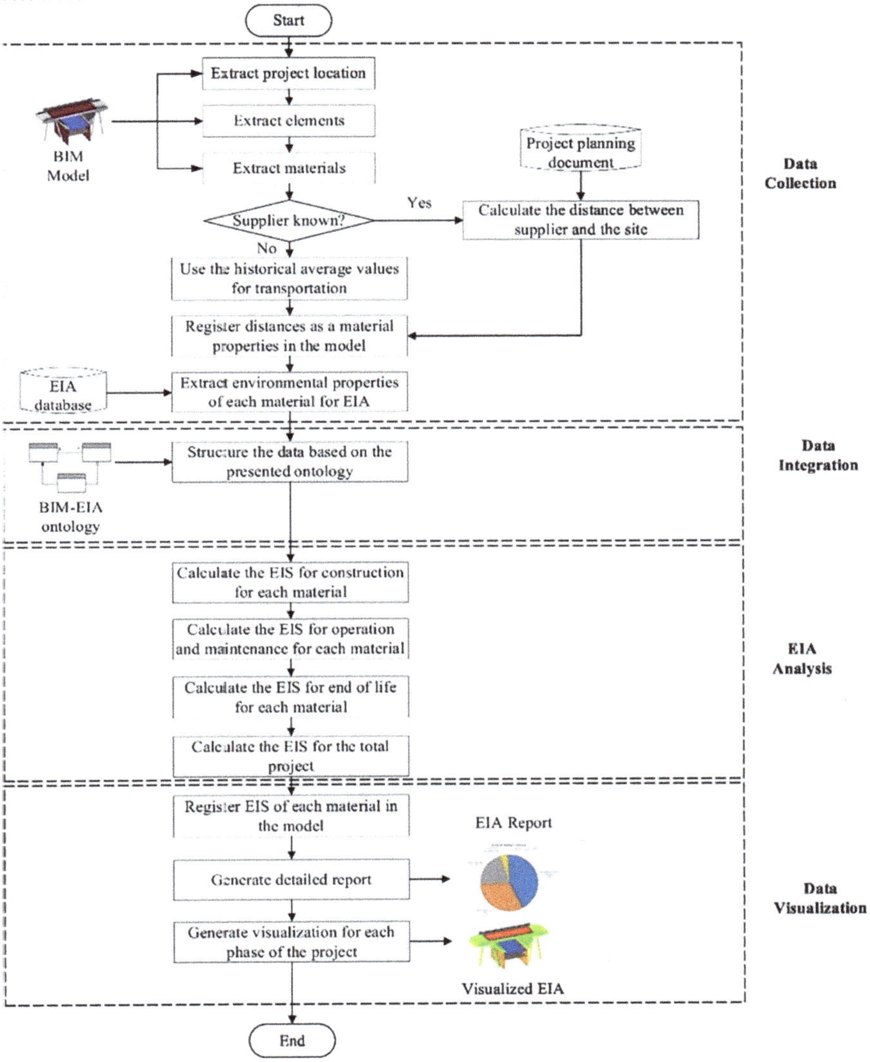

Figure 15. Different environmental impacts categories considered in EIA.

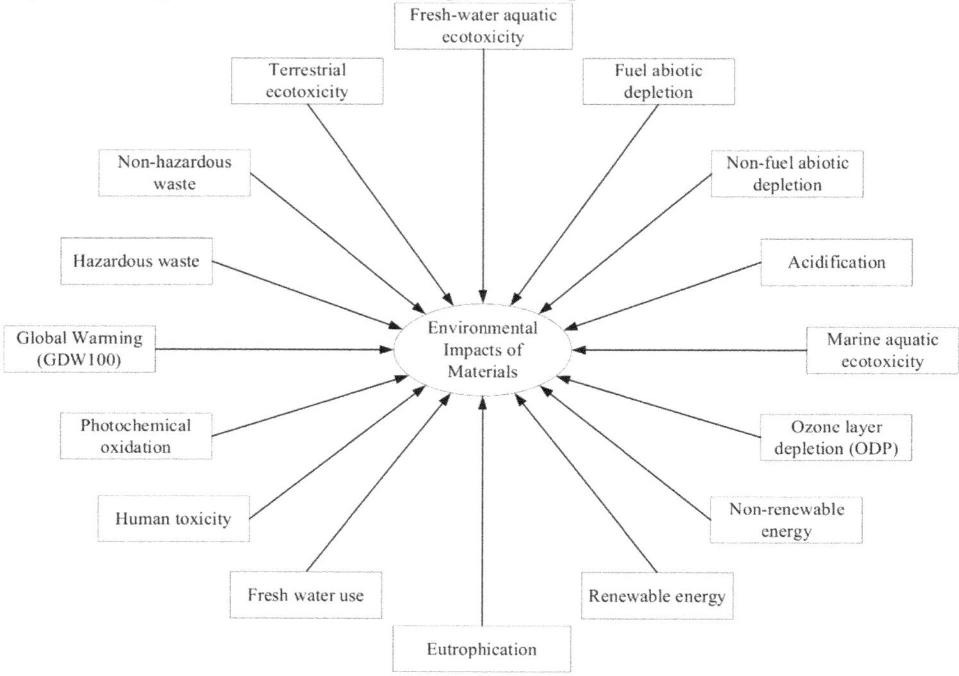

Since GIS can link spatial to descriptive information and overlap these, it is an effective tool to address such planning topics(Schaller, 1992). Single issue indicators, such as the embodied energy (EE) or carbon footprint (CF), multi-variable ones such as emergy, ecological footprint(Tab 1), CML or the eco-indicator available to measure ARE are established to evaluate the environmental load of construction sector. In this way, the environmental assessment usually involves construction materials and energy consumption, greenhouse gas (GHG) emissions or construction and demolition waste (CDW) generation. Evaluation of the environmental impacts based on databases costs of construction, and visualization of projects in a BIM environment to facilitate their insertion in urbanization project (Marrero et al., 2020).

Figure 16. Reliability of the economic and environmental assessment of construction projects (Marrero et al., 2020).

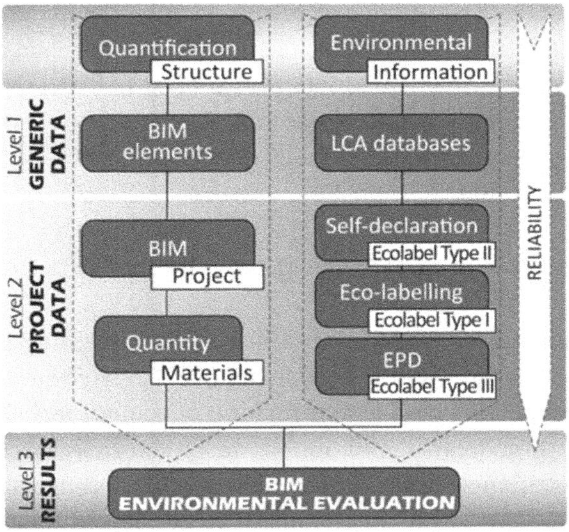

Construction cost databases, BIM environment for assisting the application of these simulations in urbanisation projects (Fernández-Rodríguez et al., 2018). Another area of interest in aerobiology is environmental impact assessment, related to the changing climate; modelling of environmental systems; life cycle analysis and sustainability assessment; characterization, management and treatment atmospheric pollution and urban mtabolism, which is a field known as Urban ecology or ecological urbanism.

Green Infrastructure Planning and Management

Therefore, the Saudi government created the Saudi Green Building Council (SGBC) in 2010 to foster the integration of green buildings within construction projects(Banani et al., 2016). By facility and system, UGI means a network of natural and engineered ecological systems localized at landscape scale that is integrated into the built environment. Green infrastructureLandscape based features such as forests, wetlands and streams that provide multiple benefits for the environment. UGI provides multiple options for 764 integration between urban development, nature conservation and promotion of public health" and thus likely to provide variety of ecosystem services from urban areas. Let´s say the green areas that help to buffer floods events or the rural peri-urban farmlands providing cities inhabitants with food and recreational services. Green infrastructure is a key element related to all of the challenges exposed before, as it represents a collection of nature and semi-natural features with associated adaptive management solutions that can both deliver a wide variety of ecosystems services. Monteiro et al., 2020 exploited the concepts that green infrastructure has to respect in order to achieve sustainable and resilient projects at the local level (Monteiro et al., 2020). Crucial pillars of UGI are living entities- plants, animals, and organisms (Source Of Life) arranged in diverse tangible forms (e.g., urban plazas & squares, street tree lines, parks and horticultural gardens) (Hanna & Comín, 2021).

Sustainable land use and zoning optimization

Using a Mathematical Framework to Improve Sustainability in Zoning, Space Utilization and Location of Facilities within Smart Cities. Recent study of (Hammad et al., 2019) presented the possible contribution of a smart technology in promoting sustainability to address demolition waste in construction. Sustainable land use patterns are location-specific and can be modeled and managed in geospatial analysis integrated with GISm (Yao et al., 2019). Land-use planning, a complex task demanding the reconciliation of various conflicting social, economic and environment factors is required worldwide including Africa (Lubida et al., 2019). Heuristically optimized land-use allocation using nondominated sorting genetic algorithm-II (NSGA-II) to maximize the four dimensions of compactness, floor area ratio (FAR), locational quality, and economic return as well as mixed use. A GIS-based multicriteria decision making (GIS-MCDM) process for optimizing the siting of a new urban residential development, assessed through sustainability dimensions social, economic and environmental benefits (Lubida et al., 2019).

Figure 17. Land-use optimization process using the MCDM approach (Rahman & Szabó, 2022).

INFRASTRUCTURE AND UTILITY MANAGEMENT

Energy-water systems integration, resource-efficient transportation and infrastructure development, and effective monitoring and management of energy use, emissions, climate risks to, and security of critical energy-water- transport-food system (Sperling & Ramaswami, 2018). Multiplier incentivesCross-scale governance arrangements The sustainable provision of water and sanitation services into the future will require advanced water supply and wastewater infrastructure asset management (IAM) to ensure satisfactory levels of service are maintained, particularly in terms of secure delivery of high-quality drinking water, flood protection due to urbanization trends, minimising drawdown on natural resources and avoiding pollution issues (Cardoso et al., 2012).

Integrated Urban Water Management (IUWM), wellknown through the works of World Bank, CSIRO, Global Water Partnership and SWITCH project. IUWM plans can be characterized as planning larger scale decentralized water supply, sewerage, drainage, liveability and ecosystem services in unison to deliver the best engineering solutions for infrastructure over the long term (Furlong et al., 2017). Now, in a time of rising planning uncertainty — such as the contemporary state of urban infrastructure today — it is harder and harder to plan processually along the linear model of decision-making following rational choice logic. Building Information Modelling (BIM) has been very popular in digital built Asset Management (AM), and the integration of Building Information Modelling (BIM) &Geographical Information Systems (GIS) are becoming more trending in this sector where it promises for better information management process control and quality delivery of functionalities.

GIS-DRIVEN URBAN ANALYTICS FOR SUSTAINABILITY

GIS based analysis of the urban sustainability grading as defined in the chosen model, validity and reliability of spatial data collected,specific type of selected GIS analysis mouthed together spectral information obtained. Geographic Information System (GIS) analysis was carried out for urban sustainability using GIS including Waste Productivity, some effects of greenhouse gases represented by the road polygons quantified by closeness to dwelling areas, industrial area as indicated from the classified satellite images land use land cover thematic maps and then vectors were created to play a role in defining their function within the scenarios (Al Salmi & Abdulmuttalib, 2012). According to (Lewin, 2012). Smart Growth, New Urbanism and the green building movement are three reform movements that identify sustainability in terms of levels of measurable data that can be easily linked to one another.

The cities or counties are located in Inner Mongolia; China were selected based on many environmental and sustainable development problems. This case section identifies two types of SUS: (1) several large size cities with developed economy, diversified industry and sufficient transportation, living condition, environment hygiene and cleaners; (2) a few small counties with low population density districts of county areas, little city construction area as rural/urban separation, extensive natural grassland but dominated primary pastoralism (Xie et al., 2022).

PUBLIC PARTICIPATION AND STAKEHOLDER ENGAGEMENT

Visualization and Simulation Tools for Community Involvement

The visualizations are first-person-view virtual environments developed with ArcGIS and Trimble SketchUp using Adobe Photoshop, and the Unity3D game engine. Ultimately, locales in turn can employ models to help clarify a host of outcomes related to community plans and strategies, including impacts on transportation efficiency, land use efficiency, energy consumption or conservation, water consumption or conservation, and greenhouse gas emissions (Newell et al., 2021).

Figure 18. Demonstrates the progression of landscape visualization tools through time.

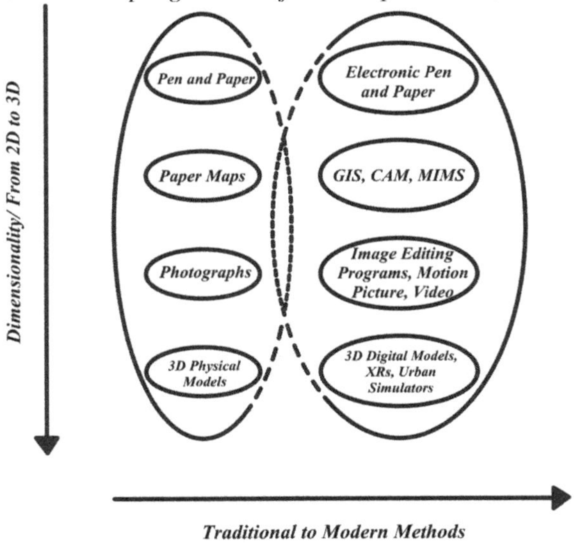

Note: GIS: Geographic Information Systems; CAM: computer-aided mapping; MIMS: Mapping Information Systems

The applications run the extent of intuitive GIS-based scenario analysis tools (e.g., PGIS PUBinPLAN CommunityViz WhatIf) to computer-based surface modeling applications (e.g., Photoshop, Autodesk Civil 3D or Unity Revit Explor Viz; and SketchUp) and more traditional approaches such as GIS maps, virtual globe, Google Earth, and so forth (Nasr-Azadani et al., 2022). Many visualizations are ineffectual to the stakeholders interactive visualizations provide a partial solution In addition, the variety of requests and also those inputs like combined on site/online variants requires an integrated use of display forms (Postert et al., 2022).

Enhancing Transparency and Collaboration Through Integrated Platforms

The Importance of Stakeholder Engagement and Transparency in Policymaking Stormont has come under the spotlight, again. Citizen, community-based and business organizations should be engaged to ensure all the appropriate perspectives and needs are taken into consideration. Transparency, in its most general definition is the public accessibility of information but specifically it is an approach to governing more than simple acts of governance: it functions to establish that transparency among other open attributes like participation and accountability is part of democratic governance. The use of open government data might improve transparency and trust in government but an idealistic vision for transparent democratic governance requires to address a number of socio-technical design challenges (Hogan et al., 2017).

Case Examples of Stakeholder-Driven Sustainable Urban Projects

Complimenting the VC process with stakeholder engagement is regarded as a potentially transformative mechanism that offers far more than just another tax venture, and is therefore suitable for generating supportive momentum towards sustainable transit-oriented community development objectives. Delineative democracy techniques offer possibilities where in the decisions taken through community engagement can get most popular mandate from each key stakeholder and mostly from the common man on any policy direction (Jillella et al., 2015). A case study focusing on Stakeholder engagement in articulating mutual vision for Suede, Dublin Region. This sense of coownership of the project was developed through a series of interviews and workshops including both top down indicators suites combined with bottom-up process measures (Siddall et al., 2013). The idea of stakeholder engagement: a frequent concept to investigated means some organizational activities, including value creation or social and sustainable innovation (Kujala et al., 2022). SEAPs which has been identified in the urban energy planning process described above (Ouhajjou et al., 2017). Stakeholder involvement in sustainability initiatives "considerably improve the quality of environmental and sustainability decisions, because it brings a broader range of information to bear on the decision-making process.

CHALLENGES AND LIMITATIONS

The fusion of GIS and BIM can be classified in to three levels namely data level, process level, and application layer. BIM has a higher richer content than GIS and it may have significantly more data than what GIS can handle natively. To build a complete integration system, differences could be stored and represented using a reference ontology as part of semantic web technologies (X. Liu et al., 2017). Scalable advanced technology, addressing urgent urban living issues — bettering the environment, budgets and lifestyle + more. IoT refers to the interconnectivity of numerous devices for remote or online monitoring, used to collect a wide variety of data on urban activities. In addition, AI could be used in analyzing data taken from many parts of the city (roads, the way commute, and number of passengers) to plan for future or to deploy various kinds of transportation schemes within the city (Ahmad et al., 2022). The accumulation, exchange and storage as well as the processing of huge amounts of such data towards serving the citizens' needs will result for sure in showing behavioral patterns, lifestyle patterns or interests or consumption habits which can consequently lead to an easy way to dig up private personal information about a citizen from their privacy (Panahi Rizi & Hosseini Seno, 2022). This is also another main sector in which IoT has played important role, e.g. maps, smart guidance, vehicle tracking, dynamic traffic lights, smart vehicles and roads this provides some new and wide variety of services in the Transportation field.

FUTURE DIRECTIONS AND RESEARCH OPPORTUNITIES

Sustainability assessment tools have been developed and operationalised internationally over the past two decades and applied to equals of sustainability that are relevant to building components including energy, water, waste, infrastructure etc. Despite the significance of sustainability assessment and its purpose for building without taking into account the environments that these engaged buildings contain within, isn't considered as a comprehensive alternative (Ameen et al., 2015). Therefore, optimizing the implementation of BIM-GIS for infrastructure and maximizing their acceptance needs integration with other technology IoT (Internet of Things), sensor technology. Spatial data infrastructure (SDI) or geospatial information infrastructure (GII). SDIs are essentially generic information infrastructures specialized in the handling of spatial data using geographic principles and tools. An example of the development of an infrastructure to share both GI and GT for the integration of spatial planning (Ran & Nedovic-Budic, 2016).

Figure 19. Fundamental concepts underlying the definition and development of SIPI.

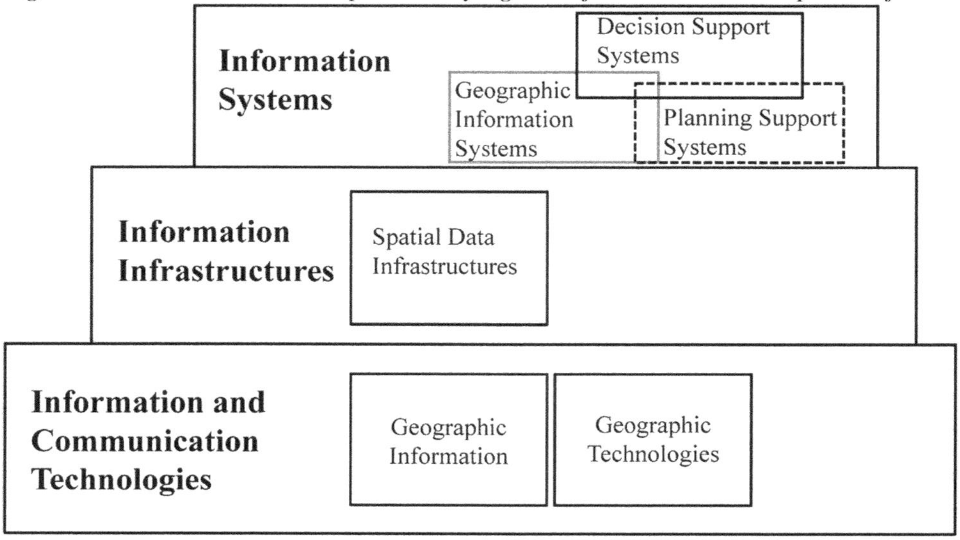

SIPI has an inherent spatial nature in the form of set of ICTs such as Geospatial Information (GI) and geospatial technology (GT). SIPI is Spatial Data Infrastructure (Ran & Nedovic-Budic, 2016) that would be integrated with spatial IS like Geographic Information System (GIS) as well as Planning Support System (PSS) and Decision Support System(DSS). AbstractSpatial decision-support tools (SDSTs) are ready to alter the way decisions are made and can have a crucial impact on the urban fabric of tomorrow. SDST in an urban planning context, using the SDSTs developed for New Zealand cities to explore: (1) how they may shape decision-making; (2) what challenges are refraining from their adequate use of spatial tools; and (3) how data, technical, and procedural issues could impact effectively adopting SDST in planning practices (Schindler et al., 2020). Spatial PSS are part of the suite of geo-information technology tools that have been largely created to assist in different stages of the planning process, such as problem diagnosis, data collection, mining and extraction spatial and temporal analysis, data modeling visualization and display scenario-building and projection plan formulation and evaluation report preparation enhanced participation facilitated through collaboration (Sudhira & Ramachandra, 2009).

As known framework a development of sustainable performance management (SPM) is suggested. The implementation of SPM prompts for a scale of performance indicators for measuring, monitoring and reporting the progress(Basu & Kumar, 2004). Experts around the world are finally acknowledging that consumers, manufacturers, governments and communities all must understand just how grave environmental risks are and worry about the consequences of climate change. China Pollution Problem and Since the manufacturing industry is one of the leading industrial waste producing as well environmental polluting sectors in the world, therefor it harms on environment sustainability (Y. Wang & Yang, 2021) studies the innovation metrics in construction and civil engineering, which are identified as some of the fields underperforming in terms of innovation compared to other areas of science and technology (Terzis, 2022). Numerous innovations centre on the economic, social and environmental pillars of sustainable development. The types of innovation that can be seen include affordable, durable, repairable, reusable, recyclable and biodegradable materials or products with costs and eco-impact reductions (Zartha Sossa et al., 2022). This is expected to decrease the input of virgin materials and generation of waste, thereby locking in or closing the "loops" or economic and ecological inflows of materials, which will make it possible for companies to develop a balanced approach for sustainable innovation via taking care of environmental, mainly economic and social targets.

CONCLUSION

Urbanization around the world is creating unprecedented challenges due to rapid population growth, degradation of the environment, scarcity in resources, and the urgent need to address climate risks. As a result, sustainable urban design and management have increasingly become critical drivers of shaping future cities that are, not only, livable but also, environmentally friendly and economically sustainable. This chapter is designed to explore how the integration of Building Information Modeling and Geographic Information Systems play a pivotal role in driving sustainable urban planning and management practices. While BIM provides granular, multidimensional models of the built environment traditionally at a building level, GIS focuses on the broad scale of the spatial context to analyze environmental, social, and infrastructural factors. The combination of these spatial intelligence technologies can support the creation of a solid foundation of spatial intelligence that would enhance urban planners' capabilities to analyze datasets, simulate sustainable alternatives, and implement decisions that would balance diverse ecological, economic, and social goals. The chapter is divided into several sections starting with laying out conceptual backgrounds and technological evolution of BIM and GIS and underlining their increasing convergence. An overview of the framework for integrating BIM and GIS is presented focused on data interoperability, standardization ideas, and key applications, challenges, and the path to integrating building-level information with the spatial context. Technological advancement, including 3D city modeling, real-time data integration, and visualization, is discussed as supporting innovative sustainable design techniques using spatial intelligence. Application concepts are later illustrated on the examples of case studies of BIM and GIS integrated work on green infrastructure, environmental impact, energy-efficient buildings, and planning optimized for land use and building. Cases depict how spatial intelligence tools could support urban infrastructure management, drive resource efficiency, and address urban environmental stressors. Following a review of the opportunity landscape, this chapter looks into the role of GIS in driving BIM and exploring more specifically spatial decision support systems, as an example, and how other GIS-driven analytics support urban sustainability, including traffic and transportation planning, and urban environmental quality monitoring. Another essential aspect is the role of the public in this process, and this chapter outlines visualization and simulation technologies as the tools that empower stakeholders and communities. Despite the promising future of the integration of BIM and GIS, this chapter also addresses existing barriers from technical obstacles, including interoperability, data quality, and the absence of unified practices. The chapter finally concludes by summarizing emerging trends such as the emergence of IoT, artificial intelligence, and digital twins that have the potential to revolutionize urban design and management further. The summary and recommendations are likely to benefit a range of stakeholders, from researchers in the field to policymakers interested in developing smarter, greener, and more resilient cities. The critical takeaway is that only the interdisciplinary nature of spatial intelligence tools will be able to meet the ambitious sustainability targets of the 21 st century urban scene.

REFERENCES

Ahmad, K., Maabreh, M., Ghaly, M., Khan, K., Qadir, J., & Al-Fuqaha, A. (2022). Developing future human-centered smart cities: Critical analysis of smart city security, Data management, and Ethical challenges. *Computer Science Review*, *43*, 100452. DOI: 10.1016/j.cosrev.2021.100452

Al Salmi, H., & Abdulmuttalib, H. M. (2012). ROLE ASSESSMENT OF GIS ANALYSIS AND ITS RELIABILITY WHILE RANKING URBAN SUSTAINABILITY USING SCENARIOS SPECIFIC TO REGIONAL CLIMATE, COMMUNITY AND CULTURE. *The International Archives of the Photogrammetry. Remote Sensing and Spatial Information Sciences, XXXIX*, *B4*, 321–326. DOI: 10.5194/isprsarchives-XXXIX-B4-321-2012

Al-Shihri, F. S. (2013). PRINCIPLES OF SUSTAINABLE DEVELOPMENT AND THEIR APPLICATION IN URBAN PLANNING IN SAUDI ARABIA. *JES. Journal of Engineering Sciences*, *41*(4), 1703–1727. DOI: 10.21608/jesaun.2013.114898

Alberti, M., & Susskind, L. (1996). Managing urban sustainability: An introduction to the special issue. *Environmental Impact Assessment Review*, *16*(4–6), 213–221. DOI: 10.1016/S0195-9255(96)00070-4

Alhazmi, H., & Loy, A. C. M. (2021). A review on environmental assessment of conversion of agriculture waste to bio-energy via different thermochemical routes: Current and future trends. *Bioresource Technology Reports*, *14*, 100682. DOI: 10.1016/j.biteb.2021.100682

Ameen, R. F. M., Mourshed, M., & Li, H. (2015). A critical review of environmental assessment tools for sustainable urban design. *Environmental Impact Assessment Review*, *55*, 110–125. DOI: 10.1016/j.eiar.2015.07.006

Amirebrahimi, S., & Rajabifard, A. (2012). AN INTEGRATED WEB-BASED 3D MODELING AND VISUALIZATION PLATFORM TO SUPPORT SUSTAINABLE CITIES. *ISPRS Annals of the Photogrammetry, Remote Sensing and Spatial Information Sciences*, *I–4*, 299–304. DOI: 10.5194/isprsannals-I-4-299-2012

Banani, R., Vahdati, M. M., Shahrestani, M., & Clements-Croome, D. (2016). The development of building assessment criteria framework for sustainable non-residential buildings in Saudi Arabia. *Sustainable Cities and Society*, *26*, 289–305. DOI: 10.1016/j.scs.2016.07.007

Basu, A. J., & Kumar, U. (2004). Innovation and Technology Driven Sustainability Performance Management Framework (ITSPM) for the Mining and Minerals Sector. *International Journal of Surface Mining. Reclamation and Environment*, *18*(2), 135–149. DOI: 10.1080/13895260412331295394

Biljecki, F., Stoter, J., Ledoux, H., Zlatanova, S., & Çöltekin, A. (2015). Applications of 3D City Models: State of the Art Review. *ISPRS International Journal of Geo-Information*, *4*(4), 2842–2889. DOI: 10.3390/ijgi4042842

Bruno, N., Rechichi, F., Achille, C., Zerbi, A., Roncella, R., & Fassi, F. (2020). INTEGRATION OF HISTORICAL GIS DATA IN A HBIM SYSTEM. *The International Archives of the Photogrammetry, Remote Sensing and Spatial Information Sciences*, *XLIII-B4-2020*, 427–434. DOI: 10.5194/isprs-archives-XLIII-B4-2020-427-2020

Bryde, D., Broquetas, M., & Volm, J. M. (2013). The project benefits of Building Information Modelling (BIM). *International Journal of Project Management*, *31*(7), 971–980. DOI: 10.1016/j.ijproman.2012.12.001

Bueno, R. E., Tetu Moysés, S., Reis Bueno, P. A., Jorge Moysés, S., De Carvalho, M. L., & Sottile França, B. H. (2013). Sustainable development and child health in the Curitiba metropolitan mesoregion, State of Paraná, Brazil. *Health & Place*, *19*, 167–173. DOI: 10.1016/j.healthplace.2012.10.007 PMID: 23262454

Buyukdemircioglu, M., & Kocaman, S. (2020). Reconstruction and Efficient Visualization of Heterogeneous 3D City Models. *Remote Sensing (Basel)*, *12*(13), 2128. DOI: 10.3390/rs12132128

Cardoso, M. A., Silva, M. S., Coelho, S. T., Almeida, M. C., & Covas, D. I. C. (2012). Urban water infrastructure asset management – a structured approach in four water utilities. *Water Science and Technology*, *66*(12), 2702–2711. DOI: 10.2166/wst.2012.509 PMID: 23109589

Dall'O', G., Zichi, A., & Torri, M. (2020). Green BIM and CIM: Sustainable Planning Using Building Information Modelling. In G. Dall'O' (Ed.), *Green Planning for Cities and Communities* (pp. 383–409). Springer International Publishing. DOI: 10.1007/978-3-030-41072-8_17

Du, J., Zou, Z., Shi, Y., & Zhao, D. (2018). Zero latency: Real-time synchronization of BIM data in virtual reality for collaborative decision-making. *Automation in Construction*, *85*, 51–64. DOI: 10.1016/j.autcon.2017.10.009

El Ammari, K., & Hammad, A. (2019). Remote interactive collaboration in facilities management using BIM-based mixed reality. *Automation in Construction*, *107*, 102940. DOI: 10.1016/j.autcon.2019.102940

ElBeheiry, D. M. (2025). An Integrated BIM and GIS Framework for Sustainable Urban Planning. *JES. Journal of Engineering Sciences*, *0*(0), 270–287. DOI: 10.21608/jesaun.2025.358173.1422

Elshaboury, N., AlMetwaly, W. M., Hesham, A., & Abbas, A. (2025). Integrated BIM-GIS framework for holistic building stock assessment using 5D geo-modeling and digital twin concepts. *Journal of Building Engineering*, *111*, 113391. DOI: 10.1016/j.jobe.2025.113391

Elsheikh, A., Alzamili, H. H., Al-Zayadi, S. K., & Alboo-Hassan, A. S. (2021). Integration of GIS and BIM in Urban Planning -A Review. *IOP Conference Series. Materials Science and Engineering*, *1090*(1), 012128. DOI: 10.1088/1757-899X/1090/1/012128

Fernández-Rodríguez, S., Cortés-Pérez, J. P., Muriel, P. P., Tormo-Molina, R., & Maya-Manzano, J. M. (2018). Environmental impact assessment of Pinaceae airborne pollen and green infrastructure using BIM. *Automation in Construction*, *96*, 494–507. DOI: 10.1016/j.autcon.2018.10.011

Furlong, C., Brotchie, R., Considine, R., Finlayson, G., & Guthrie, L. (2017). Key concepts for Integrated Urban Water Management infrastructure planning: Lessons from Melbourne. *Utilities Policy*, *45*, 84–96. DOI: 10.1016/j.jup.2017.02.004

Glander, T., & Döllner, J. (2009). Abstract representations for interactive visualization of virtual 3D city models. *Computers, Environment and Urban Systems*, *33*(5), 375–387. DOI: 10.1016/j.compenvurbsys.2009.07.003

Graziano, A. D., Ragusa, E., Trifilò, D., Triaca, L. M., Trombetti, M., & Arcidiacono, C. (2023). Interaction Between Airport Facilities And The Surrounding Area Within The GIS And BIM Interoperability. *Transportation Research Procedia*, *69*, 273–280. DOI: 10.1016/j.trpro.2023.02.172

Hammad, A. W., Akbarnezhad, A., Haddad, A., & Vazquez, E. G. (2019). Sustainable Zoning, Land-Use Allocation and Facility Location Optimisation in Smart Cities. *Energies*, *12*(7), 1318. DOI: 10.3390/en12071318

Hanna, E., & Comín, F. A. (2021). Urban Green Infrastructure and Sustainable Development: A Review. *Sustainability (Basel)*, *13*(20), 11498. DOI: 10.3390/su132011498

He, S., Zhu, D., Chen, Y., Liu, X., Chen, Y., & Wang, X. (2020). Application and problems of emergy evaluation: A systemic review based on bibliometric and content analysis methods. *Ecological Indicators*, *114*, 106304. DOI: 10.1016/j.ecolind.2020.106304

Hogan, M., Ojo, A., Harney, O., Ruijer, E., Meijer, A., Andriessen, J., Pardijs, M., Boscolo, P., Palmisano, E., Satta, M., Groff, J., Baker, M., Détienne, F., Porwol, L., Scarano, V., & Malandrino, D. (2017). Governance, Transparency and the Collaborative Design of Open Data Collaboration Platforms: Understanding Barriers, Options, and Needs. In Ojo, A., & Millard, J. (Eds.), *Government 3.0 – Next Generation Government Technology Infrastructure and Services* (Vol. 32, pp. 299–332). Springer International Publishing., DOI: 10.1007/978-3-319-63743-3_12

Ilesanmi, A. O. (2010). Urban sustainability in the context of Lagos mega-city. *Journal of Geography and Regional Planning*, *3*(10), 240–252.

Jain, M., Khan, S. A., Sharma, K., Jadhao, P. R., Pant, K. K., Ziora, Z. M., & Blaskovich, M. A. T. (2022). Current perspective of innovative strategies for bioremediation of organic pollutants from wastewater. *Bioresource Technology*, *344*, 126305. DOI: 10.1016/j.biortech.2021.126305 PMID: 34752892

Jalaei, F. (2015). *Integrate Building Information Modeling (BIM) and Sustainable Design at the Conceptual Stage of Building Projects*. DOI: 10.20381/RUOR-4260

Jillella, S., Matan, A., & Newman, P. (2015). Participatory Sustainability Approach to Value Capture-Based Urban Rail Financing in India through Deliberated Stakeholder Engagement. *Sustainability (Basel)*, *7*(7), 8091–8115. DOI: 10.3390/su7078091

Johansson, T., Segerstedt, E., Olofsson, T., & Jakobsson, M. (2016). Revealing Social Values by 3D City Visualization in City Transformations. *Sustainability (Basel)*, *8*(2), 195. DOI: 10.3390/su8020195

Kang, T. W., & Hong, C. H. (2015). A study on software architecture for effective BIM/GIS-based facility management data integration. *Automation in Construction*, *54*, 25–38. DOI: 10.1016/j.autcon.2015.03.019

Keivani, R. (2009). A review of the main challenges to urban sustainability. *International Journal of Urban Sustainable Development*, *1*(1–2), 5–16. DOI: 10.1080/19463131003704213

Kujala, J., Heikkinen, A., Jokinen, A., & Tapaninaho, R. (2022). Stakeholder Engagement in the Generation of Urban Ecosystem Services: The Case of the Vuores Stormwater System. *South Asian Journal of Business and Management Cases*, *11*(1), 49–63. DOI: 10.1177/22779779221079484

Lewin, S. S. (2012). URBAN SUSTAINABILITY AND URBAN FORM METRICS. *Journal of Green Building*, *7*(2), 44–63. DOI: 10.3992/jgb.7.2.44

Leyzerova, A., Sharovarova, E., & Alekhin, V. (2016). Sustainable Strategies of Urban Planning. *Procedia Engineering, 150*, 2055–2061. DOI: 10.1016/j.proeng.2016.07.299

Li, N., Chan, D., Mao, Q., Hsu, K., & Fu, Z. (2018). Urban sustainability education: Challenges and pedagogical experiments. *Habitat International, 71*, 70–80. DOI: 10.1016/j.habitatint.2017.11.012

Liu, J.-M., Setiazi, H., & So, P.-Y. (2023). Fisheries hydroacoustic assessment: A bibliometric analysis and direction for future research towards a blue economy. *Regional Studies in Marine Science, 60*, 102838. DOI: 10.1016/j.rsma.2023.102838

Liu, X., Wang, X., Wright, G., Cheng, J., Li, X., & Liu, R. (2017). A State-of-the-Art Review on the Integration of Building Information Modeling (BIM) and Geographic Information System (GIS). *ISPRS International Journal of Geo-Information, 6*(2), 53. DOI: 10.3390/ijgi6020053

Lubida, A., Veysipanah, M., Pilesjo, P., & Mansourian, A. (2019). LAND-USE PLANNING FOR SUSTAINABLE URBAN DEVELOPMENT IN AFRICA: A SPATIAL AND MULTI-OBJECTIVE OPTIMIZATION APPROACH. *Geodesy and Cartography (Vilnius), 45*(5), 1–15. DOI: 10.3846/gac.2019.6691

Luederitz, C., Lang, D. J., & Von Wehrden, H. (2013). A systematic review of guiding principles for sustainable urban neighborhood development. *Landscape and Urban Planning, 118*, 40–52. DOI: 10.1016/j.landurbplan.2013.06.002

Marrero, M., Wojtasiewicz, M., Martínez-Rocamora, A., Solís-Guzmán, J., & Alba-Rodríguez, M. D. (2020). BIM-LCA Integration for the Environmental Impact Assessment of the Urbanization Process. *Sustainability (Basel), 12*(10), 4196. DOI: 10.3390/su12104196

Marzouk, M., & Othman, A. (2020). Planning utility infrastructure requirements for smart cities using the integration between BIM and GIS. *Sustainable Cities and Society, 57*, 102120. DOI: 10.1016/j.scs.2020.102120

Milojević, B., & Kuvač, I. (2023). Recognizing Principles of Integrated Urban Planning in Historical Development of the City: A Case Study of Banja Luka. *Journal of Urban History, 49*(2), 290–308. DOI: 10.1177/00961442211030624

Monteiro, R., Ferreira, J., & Antunes, P. (2020). Green Infrastructure Planning Principles: An Integrated Literature Review. *Land (Basel), 9*(12), 525. DOI: 10.3390/land9120525

Nasr-Azadani, E., Wardrop, D., & Brooks, R. (2022). Is the rapid development of visualization techniques enhancing the quality of public participation in natural resource policy and management? A systematic review. *Landscape and Urban Planning*, *228*, 104586. DOI: 10.1016/j.landurbplan.2022.104586

Newell, R., McCarthy, N., Picketts, I., Davis, F., Hovem, G., & Navarrete, S. (2021). Communicating complexity: Interactive model explorers and immersive visualizations as tools for local planning and community engagement. *Facets*, *6*, 287–316. DOI: 10.1139/facets-2020-0045

Ouhajjou, N., Loibl, W., Fenz, S., & Tjoa, A. M. (2017). Stakeholder-oriented energy planning support in cities. *Sustainable Cities and Society*, *28*, 482–492. DOI: 10.1016/j.scs.2016.08.006

Panahi Rizi, M. H., & Hosseini Seno, S. A. (2022). A systematic review of technologies and solutions to improve security and privacy protection of citizens in the smart city. *Internet of Things : Engineering Cyber Physical Human Systems*, *20*, 100584. DOI: 10.1016/j.iot.2022.100584

Paul, S. (2018). Geospatial world. *Urban Planning and GIS*. https://geospatialworld.net/blogs/gis-in-urban-planning/

Pavelka, K., Matoušková, E., & Pavelka, K.Jr. (2019). THE CONTRIBUTION OF GEOMATIC TECHNOLOGIES TO BIM. *The International Archives of the Photogrammetry, Remote Sensing and Spatial Information Sciences*, *XLII-5*(W3), 85–89. DOI: 10.5194/isprs-archives-XLII-5-W3-85-2019

Postert, P., Wolf, A. E. M., & Schiewe, J. (2022). Integrating Visualization and Interaction Tools for Enhancing Collaboration in Different Public Participation Settings. *ISPRS International Journal of Geo-Information*, *11*(3), 156. DOI: 10.3390/ijgi11030156

Qiu, Y., Zuo, S., Yu, Z., Zhan, Y., & Ren, Y. (2021). Discovering the effects of integrated green space air regulation on human health: A bibliometric and meta-analysis. *Ecological Indicators*, *132*, 108292. DOI: 10.1016/j.ecolind.2021.108292

Rahman, M., & Szabó, G. (2022). Sustainable Urban Land-Use Optimization Using GIS-Based Multicriteria Decision-Making (GIS-MCDM) Approach. *ISPRS International Journal of Geo-Information*, *11*(5), 313. DOI: 10.3390/ijgi11050313

Ran, J., & Nedovic-Budic, Z. (2016). Integrating spatial planning and flood risk management: A new conceptual framework for the spatially integrated policy infrastructure. *Computers, Environment and Urban Systems*, *57*, 68–79. DOI: 10.1016/j.compenvurbsys.2016.01.008

Rodrigues, F., Teixeira, J., Matos, R., & Rodrigues, H. (2019). Development of a Web Application for Historical Building Management through BIM Technology. *Advances in Civil Engineering*, *2019*(1), 9872736. DOI: 10.1155/2019/9872736

Saccucci, M., & Pelliccio, A. (2018). Integrated BIM-GIS system for the enhancement of urban heritage. *2018 Metrology for Archaeology and Cultural Heritage (MetroArchaeo)*, 222–226. DOI: 10.1109/MetroArchaeo43810.2018.13625

Schaller, J. (1992). GIS application in environmental planning and assessment. *Computers, Environment and Urban Systems*, *16*(4), 337–353. DOI: 10.1016/0198-9715(92)90015-J

Schindler, M., Dionisio, R., & Kingham, S. (2020). Challenges of Spatial Decision-Support Tools in Urban Planning: Lessons from New Zealand's Cities. *Journal of Urban Planning and Development*, *146*(2), 04020012. DOI: 10.1061/(ASCE)UP.1943-5444.0000575

Sheina, S., Chubarova, K., Dementeev, D., & Kalitkin, A. (2023). Integration of BIM and GIS Technologies for Sustainable Development of the Construction Industry. In Guda, A. (Ed.), *Networked Control Systems for Connected and Automated Vehicles* (Vol. 509, pp. 1303–1311). Springer International Publishing., DOI: 10.1007/978-3-031-11058-0_132

Sheth, A., & Kochut, K. J. (1998). Workflow Applications to Research Agenda: Scalable and Dynamic Work Coordination and Collaboration Systems. In Doğaç, A., Kalinichenko, L., Özsu, M. T., & Sheth, A. (Eds.), *Workflow Management Systems and Interoperability* (pp. 35–60). Springer Berlin Heidelberg., DOI: 10.1007/978-3-642-58908-9_3

Shiode, N. (2000). 3D urban models: Recent developments in the digital modelling of urban environments in three-dimensions. *GeoJournal*, *52*(3), 263–269. DOI: 10.1023/A:1014276309416

Siddall, E., Grey, T., & Dyer, M. (2013). Indicators and stakeholder engagement: A Dublin case study. *Proceedings of the Institution of Civil Engineers. Engineering Sustainability*, *166*(2), 85–97. DOI: 10.1680/ensu.12.00004

Sperling, J. B., & Ramaswami, A. (2018). Cities and "budget-based" management of the energy-water-climate nexus: Case studies in transportation policy, infrastructure systems, and urban utility risk management. *Environmental Progress & Sustainable Energy*, *37*(1), 91–107. DOI: 10.1002/ep.12765

Sudhira, H. S., & Ramachandra, T. V. (2009). A Spatial Planning Support System for Managing Bangalore's Urban Sprawl. In Geertman, S., & Stillwell, J. (Eds.), *Planning Support Systems Best Practice and New Methods* (Vol. 95, pp. 175–190). Springer Netherlands., DOI: 10.1007/978-1-4020-8952-7_9

Sun, Y., & Li, S. (2016). Real-time collaborative GIS: A technological review. *ISPRS Journal of Photogrammetry and Remote Sensing*, *115*, 143–152. DOI: 10.1016/j.isprsjprs.2015.09.011

Terzis, D. (2022). Monitoring innovation metrics in construction and civil engineering: Trends, drivers and laggards. *Developments in the Built Environment*, *9*, 100064. DOI: 10.1016/j.dibe.2021.100064

Tsilimantou, E., Delegou, E. T., Nikitakos, I. A., Ioannidis, C., & Moropoulou, A. (2020). GIS and BIM as Integrated Digital Environments for Modeling and Monitoring of Historic Buildings. *Applied Sciences (Basel, Switzerland)*, *10*(3), 1078. DOI: 10.3390/app10031078

Verma, P., & Raghubanshi, A. S. (2018). Urban sustainability indicators: Challenges and opportunities. *Ecological Indicators*, *93*, 282–291. DOI: 10.1016/j.ecolind.2018.05.007

Wang, H., Pan, Y., & Luo, X. (2019). Integration of BIM and GIS in sustainable built environment: A review and bibliometric analysis. *Automation in Construction*, *103*, 41–52. DOI: 10.1016/j.autcon.2019.03.005

Wang, M., Deng, Y., Won, J., & Cheng, J. C. P. (2019). An integrated underground utility management and decision support based on BIM and GIS. *Automation in Construction*, *107*, 102931. DOI: 10.1016/j.autcon.2019.102931

Wang, Y., & Yang, Y. (2021). Analyzing the green innovation practices based on sustainability performance indicators: A Chinese manufacturing industry case. *Environmental Science and Pollution Research International*, *28*(1), 1181–1203. DOI: 10.1007/s11356-020-10531-7 PMID: 32833175

Xia, H., Liu, Z., Efremochkina, M., Liu, X., & Lin, C. (2022). Study on city digital twin technologies for sustainable smart city design: A review and bibliometric analysis of geographic information system and building information modeling integration. *Sustainable Cities and Society*, *84*, 104009. DOI: 10.1016/j.scs.2022.104009

Xie, Y., Liu, C., Chang, S., & Jiang, B. (2022). Urban Sustainability: Integrating Socioeconomic and Environmental Data for Multi-Objective Assessment. *Sustainability (Basel)*, *14*(15), 9142. DOI: 10.3390/su14159142

Xu, Z., & Coors, V. (2012). Combining system dynamics model, GIS and 3D visualization in sustainability assessment of urban residential development. *Building and Environment*, *47*, 272–287. DOI: 10.1016/j.buildenv.2011.07.012

Yamamura, S., Fan, L., & Suzuki, Y. (2017). Assessment of Urban Energy Performance through Integration of BIM and GIS for Smart City Planning. *Procedia Engineering*, *180*, 1462–1472. DOI: 10.1016/j.proeng.2017.04.309

Yao, J., Murray, A. T., Wang, J., & Zhang, X. (2019). Evaluation and development of sustainable urban land use plans through spatial optimization. *Transactions in GIS*, *23*(4), 705–725. DOI: 10.1111/tgis.12531

Yigitcanlar, T., & Teriman, S. (2015). Rethinking sustainable urban development: Towards an integrated planning and development process. *International Journal of Environmental Science and Technology*, *12*(1), 341–352. DOI: 10.1007/s13762-013-0491-x

Zartha Sossa, J. W., Gaviria Suárez, J. F., López Suárez, N. M., Rebolledo, J. L. S., Orozco Mendoza, G. L., & Vélez Suárez, V. (2022). Innovation Systems and Sustainability. Development of a Methodology on Innovation Systems for the Measurement of Sustainability Indicators in Regions Based on a Colombian Case Study. *Sustainability (Basel)*, *14*(23), 15955. DOI: 10.3390/su142315955

Chapter 8

Use of BIM and GIS in Maintenance and Operation of Urban Infrastructure

P. Selvakumar
https://orcid.org/0000-0002-3650-4548
Department of Science and Humanities, Nehru Institute of Technology, Coimbatore, India

ABSTRACT

Urban infrastructure is the backbone of modern cities, encompassing transportation systems, utilities, buildings, public spaces, and communication networks. As cities grow and become more complex, the need for intelligent, data-driven infrastructure planning and management becomes crucial. Two powerful technologies—Building Information Modeling (BIM) and Geographic Information Systems (GIS)—are revolutionizing the way urban infrastructure is designed, constructed, operated, and maintained. By integrating these systems, urban planners, engineers, architects, and policymakers can make more informed decisions, optimize resources, and improve the quality and sustainability of urban environments. Building Information Modeling (BIM) is a digital representation of the physical and functional characteristics of a facility. Unlike traditional 2D drawings, BIM uses 3D models enriched with data that can include geometry, spatial relationships.

DOI: 10.4018/979-8-3373-4058-6.ch008

INTRODUCTION TO BIM AND GIS

Urban infrastructure is the backbone of modern cities, encompassing transportation systems, utilities, buildings, public spaces, and communication networks. As cities grow and become more complex, the need for intelligent, data-driven infrastructure planning and management becomes crucial. Two powerful technologies—Building Information Modeling (BIM) and Geographic Information Systems (GIS)—are revolutionizing the way urban infrastructure is designed, constructed, operated, and maintained. By integrating these systems, urban planners, engineers, architects, and policymakers can make more informed decisions, optimize resources, and improve the quality and sustainability of urban environments. Building Information Modeling (BIM) is a digital representation of the physical and functional characteristics of a facility. Unlike traditional 2D drawings, BIM uses 3D models enriched with data that can include geometry, spatial relationships.Urban infrastructure is the backbone of modern cities, encompassing transportation systems, utilities, buildings, public spaces, and communication networks. As cities grow and become more complex, the need for intelligent, data-driven infrastructure planning and management becomes crucial. (Cepa J. J. et al. 2024) Two powerful technologies—Building Information Modeling (BIM) and Geographic Information Systems (GIS)—are revolutionizing the way urban infrastructure is designed, constructed, operated, and maintained. By integrating these systems, urban planners, engineers, architects, and policymakers can make more informed decisions, optimize resources, and improve the quality and sustainability of urban environments. (Sreekanthaswamy et al. 2025) Building Information Modeling (BIM) is a digital representation of the physical and functional characteristics of a facility. Unlike traditional 2D drawings, BIM uses 3D models enriched with data that can include geometry, spatial relationships, geographic information, quantities, and properties of building components. BIM facilitates collaboration among stakeholders throughout the lifecycle of a project—from conceptual design to construction, operation, and even demolition. BIM models are developed using parametric design tools that enable real-time modifications and ensure consistency across architectural, structural, and MEP (Mechanical, Electrical, and Plumbing) systems. (Sreekanthaswamy et al. 2025) This results in fewer errors, reduced rework, and enhanced efficiency. Moreover, BIM supports the concept of lifecycle management, where asset information can be continuously updated and used for maintenance, renovations, and retrofitting. With the incorporation of 4D (time), 5D (cost), and 6D (sustainability) dimensions, BIM enables a holistic approach to managing infrastructure over time.

Geographic Information Systems (GIS), on the other hand, are designed to capture, store, manipulate, analyze, manage, and visualize spatial or geographic data. GIS allows users to understand patterns, relationships, and trends in the form of maps, charts, and reports. It is widely used in urban planning, transportation, environmental management, public health, and emergency response. (Kommineni M. et al. 2025) GIS can manage large-scale spatial data and perform complex analyses, such as land suitability, environmental impact, route optimization, and risk assessment. Unlike BIM, which focuses on the detailed design and management of individual buildings or infrastructure elements, GIS operates at a larger scale—providing insights into how these elements interact within the broader urban and regional context. For instance, GIS can analyze how the location of a new transportation hub affects traffic patterns, pollution levels, and accessibility across a city. It helps in identifying high-risk zones, planning for climate resilience, and optimizing public services distribution. (Cepa J. J. et al. 2024) The Integration of BIM and GIS represents a paradigm shift in infrastructure management. While each system excels in its domain, their combination provides a comprehensive framework for spatial and structural analysis. Integrating BIM with GIS allows the micro-level detail of buildings to be embedded into macro-level geographic contexts. This fusion enables seamless data exchange, improved interoperability, and enhanced decision-making for urban infrastructure development. For example, integrating BIM and GIS allows urban planners to place a BIM-modeled building within a real-world GIS environment, considering terrain, zoning regulations, flood zones, and transportation networks. (Kommineni M. et al. 2025) This holistic view improves site selection, regulatory compliance, and environmental impact assessments. Furthermore, emergency responders can benefit from integrated models by understanding building layouts in relation to city infrastructure, improving response times and effectiveness during crises.

The Role of BIM and GIS in Urban Infrastructure Management

Maintenance and operation (M&O) are critical yet often underemphasized components of urban infrastructure management. (Pelden et al. 2025) While the planning, design, and construction of infrastructure systems receive significant attention and investment, it is the post-construction phase—operation and maintenance—that ultimately determines the performance, longevity, and sustainability of urban assets. Properly maintained infrastructure ensures the safety, reliability, and efficiency of cities, contributing directly to economic development, public health, environmental quality, and the overall quality of life. Urban infrastructure includes a wide range of assets such as roads, bridges, water supply systems, sewage networks, (Tamilarasi C. et al. 2025) electrical grids, public buildings, and transportation facilities. These assets are subject to wear and tear from usage, environmental conditions, aging, and unexpected events such as natural disasters. Without regular maintenance and efficient operation, infrastructure deteriorates, leading to service disruptions, increased repair costs, safety hazards, and in extreme cases, catastrophic failures. The operation of infrastructure involves the day-to-day management and functioning of systems to ensure they perform as intended. This includes activities such as monitoring traffic signals, managing water flow in treatment plants, operating public transport systems, and ensuring the proper functioning of HVAC systems in buildings. Maintenance, on the other hand, involves the routine and corrective activities required to keep infrastructure in good working condition. (Tamilarasi C. et al. 2025) This can range from cleaning drainage systems and repairing potholes to overhauling aging electrical components and reinforcing structural elements of bridges.

One of the most compelling reasons for prioritizing maintenance and operation is **cost-**effectiveness. Deferred maintenance often leads to more severe degradation that requires extensive and expensive repairs or complete asset replacement. Studies have shown that every dollar spent on preventive maintenance can save several dollars in future repair costs. Furthermore, well-maintained infrastructure supports economic productivity by minimizing downtime and ensuring the efficient movement of goods, services, and people. (Limbore N. V. et al. 2025) Safety is another paramount concern. Poorly maintained roads increase the risk of accidents, faulty water systems can cause contamination, and neglected electrical infrastructure poses fire hazards. The tragic collapse of bridges or buildings due to overlooked maintenance is a stark reminder of the importance of proactive asset management. (Pelden S. et al. 2025) Regular inspections, condition assessments, and risk-based maintenance strategies help identify vulnerabilities before they escalate into emergencies. In addition to safety and cost, service quality and user satisfaction are directly tied to maintenance and operation. Citizens expect consistent, high-quality public services—reliable electricity, clean drinking water, accessible transportation, and safe public spaces. (Limbore N. V. et al. 2025) Infrastructure that is poorly operated or inadequately maintained can lead to public dissatisfaction, decreased property values, and diminished trust in government institutions. Responsive maintenance systems that quickly address service interruptions or defects can significantly enhance public perception and community engagement.

Building Information Modeling (BIM): Digital Twins for Infrastructure

Urban infrastructure management is a complex, multifaceted domain that involves planning, designing, constructing, operating, maintaining, and upgrading a vast network of interconnected systems such as roads, bridges, water supply, sanitation, power grids, public transportation, and built environments. (Sohaib M. et al. 2024) With rapid urbanization, climate change, population growth, and aging infrastructure, cities are increasingly challenged to provide sustainable, efficient, and resilient services to their citizens. (Chib S. et al. 2025) In this evolving context, two transformative technologies—Building Information Modeling (BIM) and Geographic Information Systems (GIS)—have emerged as critical tools that empower stakeholders to manage urban infrastructure with unprecedented precision, intelligence, and integration. Building Information Modeling (BIM) is a digital representation of physical and functional characteristics of infrastructure assets. It provides detailed, object-based, three-dimensional (3D) models that integrate geometric data with attributes such as material specifications, structural loads, energy usage, and lifecycle costs. (Chib S. et al. 2025) BIM enables collaborative design, simulation, and analysis across the lifecycle of a project, from initial conceptualization to construction, facility management, and even demolition. By centralizing and visualizing data in a shared digital environment, BIM improves coordination among architects, engineers, contractors, and operators, thereby reducing design conflicts, construction errors, and maintenance uncertainties.

On the other hand, Geographic Information Systems (GIS) specialize in spatial data management and analysis. GIS allows for the capture, storage, manipulation, and visualization of geographic and environmental data tied to specific locations on Earth. (Dahiya R. et al. 2025) It excels at representing large-scale infrastructure networks within the spatial context of urban landscapes. GIS is used extensively in urban planning, land use analysis, transportation modeling, environmental monitoring, and disaster preparedness. With its ability to manage and analyze multiple layers of spatial data, GIS supports decision-making by identifying patterns, assessing risks, and modeling future scenarios. When BIM and GIS are integrated, they create a powerful synergy that combines the micro-level detail of infrastructure components with the macro-level understanding of their spatial and environmental context. (Sohaib M. et al. 2024) This integration enhances urban infrastructure management in several key ways. First, it supports holistic planning by enabling stakeholders to design infrastructure projects that align with geographical realities such as topography, land use, zoning regulations, flood zones, and existing utilities. (Dahiya R. et al. 2025) For instance, a BIM-designed subway station can be placed within a GIS-enabled city map to analyze connectivity to roads, public transport lines, and nearby amenities, ensuring better accessibility and urban integration. Second, the integration supports efficient construction management (Popov P. et al. 2025) By embedding GIS data such as soil conditions, traffic patterns, or environmental constraints into BIM models, project teams can plan construction logistics more effectively, mitigate risks, and minimize disruptions. Construction sequencing (4D BIM), cost estimation (5D BIM), and energy analysis (6D BIM) can all be enhanced through GIS layers that provide real-world data on surrounding conditions. This not only streamlines project execution but also helps in regulatory compliance and environmental impact assessments.

BIM for Maintenance Planning and Scheduling: Discussion of maintenance planning.

Building Information Modeling (BIM) has revolutionized the architecture, engineering, and construction (AEC) industries by providing a comprehensive digital framework for designing, constructing, and managing built environments. (Mushtaha A. W. et al. 2024) In the domain of infrastructure, BIM serves as a pivotal tool that enables the creation of intelligent 3D models enriched with data describing the physical and functional attributes of assets such as roads, bridges, tunnels, airports, railways, utility networks, and public buildings. (Samal A. et al. 2025) As cities face increasing pressure from population growth, urban sprawl, and climate change, BIM is being recognized not merely as a design tool, but as a foundational technology in creating digital twins—dynamic, data-driven replicas of physical infrastructure systems that mirror real-world conditions and behavior in real time. A digital twin is a virtual model of a physical asset that incorporates real-time data from sensors, historical performance records, and predictive analytics to simulate and analyze behavior, performance, and conditions. BIM forms the structural and informational core of digital twins in infrastructure. (Samal A. et al. 2025) Through its ability to integrate geometrical precision, system behavior, scheduling data (4D), cost information (5D), and performance metrics (6D and beyond), BIM creates a comprehensive digital representation that evolves alongside the physical asset. This integration allows for advanced simulation, scenario testing, maintenance planning, and optimization throughout the lifecycle of infrastructure systems.

At the design and planning stage, BIM facilitates visualization, coordination, and analysis. Infrastructure projects are inherently complex, involving numerous disciplines and stakeholders. (Mushtaha A. W. et al. 2024) Traditional 2D blueprints often result in misinterpretations and clashes, especially in multi-disciplinary projects involving structural, mechanical, electrical, and civil systems. BIM addresses this challenge through clash detection, collaborative modeling, and data interoperability, which improves design accuracy and reduces rework. (Kaur M. et al. 2025) When embedded in a digital twin framework, BIM extends beyond the design phase to include performance forecasting, environmental impact assessments, and constructability analysis, providing stakeholders with a comprehensive tool for strategic decision-making. During construction, BIM models serve as digital control centers that inform sequencing, logistics, and quality assurance. With 4D BIM, construction schedules are integrated into the 3D model, allowing teams to simulate and optimize the construction process, identify potential delays, and plan mitigation strategies. (Kaur M. et al. 2025) When enhanced by digital twin capabilities, real-time field data (e.g., from drones, RFID, or IoT sensors) can be linked to the model, enabling progress monitoring, compliance verification, and on-site issue resolution. This real-time feedback loop ensures that the digital model remains a living document, reflecting the current state of the asset throughout construction.

BIM for Facility Management: Asset Tracking and Management

Building Information Modeling (BIM) has long been recognized as a powerful tool in the design and construction phases of infrastructure projects, offering data-rich 3D models that enhance collaboration, coordination, and visualization. However, one of BIM's most transformative applications lies beyond construction—specifically in maintenance planning and scheduling. In the operational phase of infrastructure lifecycle management, BIM serves as a dynamic repository of structured asset data that facilitates proactive, efficient, and cost-effective maintenance strategies. (Gao Y. et al. 2024) The integration of BIM into facility management and maintenance workflows allows asset owners, facility managers, and operators to shift from reactive maintenance approaches to predictive and preventive models, ultimately improving asset performance, extending service life, and reducing long-term operational costs. Maintenance planning is critical for infrastructure assets such as buildings, roads, bridges, tunnels, rail systems, water supply networks, and energy grids. (Soundariya A. et al. 2025) These systems experience wear and degradation over time due to environmental exposure, mechanical use, material fatigue, and other factors. In traditional maintenance models, facilities often operate on fixed schedules or respond reactively to breakdowns, which can result in unanticipated downtime, higher repair costs, and safety risks. BIM changes this paradigm by embedding detailed, lifecycle-based information about every component of an asset into a centralized model. This includes manufacturer data, warranty information, installation dates, material properties, expected lifespans, and maintenance manuals. With this data readily accessible, maintenance teams can make informed decisions about when, where, and how to perform maintenance activities.

At the core of BIM's utility for maintenance is the concept of the digital as-built model. Unlike 2D as-builts or fragmented maintenance logs, a BIM model provides a spatially accurate and information-rich representation of the actual built environment as it exists. This model serves as a "single source of truth" that is updated throughout the project lifecycle, incorporating modifications, repairs, and operational data. (Gao Y. et al. 2024) When integrated into Computerized Maintenance Management Systems (CMMS) or Enterprise Asset Management (EAM) platforms, BIM allows for streamlined maintenance scheduling, task assignment, and performance tracking. For preventive maintenance, BIM supports the development of systematic schedules based on asset-specific data rather than generic timelines. For example, HVAC units in a commercial building might have different usage rates depending on location, exposure, or occupancy. (Soundariya A. et al. 2025) BIM allows maintenance planners to assess each unit's specifications, usage history, and environmental context to create tailored maintenance schedules that minimize downtime and maximize performance. Similarly, for infrastructure such as bridges or tunnels, BIM models can track exposure to environmental stressors—such as temperature fluctuations, traffic loads, or moisture levels—and suggest maintenance intervals based on predictive deterioration models. BIM also enables predictive maintenance through the integration of Internet of Things (IoT) sensors and real-time data analytics. These sensors can monitor conditions such as vibration, temperature, humidity, and pressure in critical components. (Van Tan N. et al. 2025) When connected to a BIM model, this sensor data can be visualized in context, allowing facility managers to monitor asset health continuously and receive alerts when conditions deviate from optimal parameters. (Van Tan N. et al. 2025) Predictive algorithms can then forecast potential failures, prioritize interventions, and suggest optimal maintenance windows, reducing the need for emergency repairs and extending asset life.

Geographic Information Systems (GIS): Spatial Analysis and Mapping

Facility management (FM) plays a vital role in the lifecycle of built infrastructure, encompassing a wide range of responsibilities such as space management, maintenance, energy use optimization, asset tracking, operational efficiency, and occupant safety. Traditionally, facility managers relied on fragmented documents, spreadsheets, and outdated as-built drawings to track assets and coordinate maintenance. (Sivakumar A. et al. 2025) These conventional methods often led to inefficiencies, data loss, redundant efforts, and increased operational costs. With the emergence and growing adoption of Building Information Modeling (BIM), facility management has undergone a paradigm shift. BIM offers a centralized, digital, and data-rich platform that not only streamlines asset tracking but also enhances overall facility operations through visualization, automation, and real-time data integration. BIM extends far beyond its original application in design and construction. In facility management, BIM serves as a dynamic digital twin of the built environment—a living model that contains precise geometric representations and detailed metadata about every building component and infrastructure asset. This includes mechanical systems, electrical networks, plumbing installations, HVAC equipment, fire safety systems, lighting fixtures, elevators, structural components, and even furniture. (Lee J. et al. 2024) Each object in a BIM model is more than just a visual element; it is a data container with critical information such as model numbers, manufacturer details, maintenance schedules, warranty data, installation dates, and expected service life.

(Lee J. et al. 2024) This embedded intelligence allows facility managers to locate, track, monitor, and maintain assets with exceptional accuracy and efficiency.

Asset tracking, one of the most fundamental responsibilities in facility management, benefits significantly from BIM's object-oriented architecture. In conventional FM workflows, locating specific assets for inspection, maintenance, or replacement often involves manual searches through physical documents or time-consuming on-site checks. BIM simplifies this process by providing a visual, spatially accurate interface that allows users to "click" on an asset in a 3D model and instantly access all relevant data. For instance, if a facility manager needs to service an air handling unit, they can locate it within the BIM model, view its operational history, check service records, verify warranty status, and schedule a technician—all from a single interface. (Sivakumar A. et al. 2025) This dramatically improves response time, reduces labor costs, and enhances overall service quality. Moreover, BIM supports mobile and field-accessible asset management, allowing maintenance teams to access real-time data via tablets or smartphones while on site. Integrated with QR codes or RFID tags, each asset can be scanned to pull up its digital profile from the BIM model. (Rotilio M. et al. 2024) This mobility not only enhances task execution but also enables technicians to update records on the fly, ensuring that asset data remains current and reliable. This real-time, bi-directional data flow supports a proactive maintenance culture, where decisions are based on live information rather than static reports. Beyond tracking physical location and attributes, BIM also enables life cycle asset management. It allows facility managers to forecast maintenance needs, plan replacements, and perform cost analysis over time. By leveraging BIM data, organizations can develop long-term capital improvement plans, prioritize critical infrastructure upgrades, and optimize resource allocation. (Rotilio M. et al. 2024) For example, a facility's BIM model can be used to simulate equipment degradation over time, helping managers determine when to repair or replace systems to avoid failures and minimize downtime. This predictive capability is particularly important in large-scale or mission-critical facilities such as hospitals, airports, government buildings, or data centers.

GIS for Infrastructure Asset Management: Mapping and Analysis

Geographic Information Systems (GIS) represent one of the most transformative technologies in modern infrastructure planning and management. At its core, GIS is a framework for gathering, managing, analyzing, and visualizing spatial and geographic data. Unlike conventional information systems that deal with tabular or linear data, GIS integrates location-based data with attribute information, allowing users to uncover patterns, relationships, and trends within spatial contexts. (Akindele

O. et al. 2025) This unique capability makes GIS indispensable in urban planning, environmental monitoring, disaster management, infrastructure asset tracking, and policy development. (Borkowski A. S. et al. 2025) Through sophisticated tools for spatial analysis and mapping, GIS empowers professionals to make informed decisions, manage resources efficiently, and understand complex interactions between human activities and natural systems. Spatial analysis is the heart of GIS functionality. It refers to the examination of geographic patterns to understand the spatial arrangement of features and the processes affecting them. (Borkowski A. S. et al. 2025) GIS enables users to conduct a wide range of spatial analyses such as proximity analysis, overlay analysis, network analysis, interpolation, and spatial statistics. These techniques allow planners and decision-makers to answer crucial questions like: Where are the underserved populations in a city? Which areas are most vulnerable to flooding? How far are emergency services from high-risk zones? By using GIS for such analyses, urban planners can optimize the location of public services, transportation routes, green spaces, and utilities, ensuring equitable and sustainable urban development.

One of the most fundamental capabilities of GIS is mapping, which translates complex datasets into visual representations of the real world. Maps created with GIS are dynamic and interactive, allowing users to zoom, filter, and analyze data layers based on specific criteria. Unlike static paper maps, GIS-based maps can be updated in real time, providing up-to-date insights into changing conditions. (Akindele O. et al. 2025) This dynamic mapping ability is particularly valuable in scenarios such as emergency response, where situational awareness must evolve rapidly. For example, during a natural disaster, GIS can display live information about road closures, shelter locations, infrastructure damage, and population movement, enabling emergency services to respond swiftly and effectively. (Piras G. et al. 2024) GIS supports data integration from diverse sources, making it a powerful platform for multi-disciplinary analysis. Satellite imagery, remote sensing data, GPS, field surveys, sensor networks, drone imagery, and administrative records can all be incorporated into GIS databases. This integration creates a holistic view of the environment, enabling the analysis of complex interrelationships. For instance, in environmental management, GIS can combine topographic maps with vegetation indices and pollution data to assess the health of ecosystems. In transportation planning, GIS can overlay demographic data with traffic flow models and land use patterns to design efficient transit systems. In urban infrastructure management, GIS plays a crucial role in asset mapping, condition monitoring, and risk assessment. Municipalities use GIS to inventory physical assets such as roads, bridges, water pipelines, electrical grids, and public buildings. Each asset can be geo-tagged and linked to attribute data such as installation date, maintenance history, material type, and condition rating. This spatial database allows asset managers to prioritize repairs, schedule maintenance, and forecast budget needs based on geographic distribution and condition trends. (Piras G. et al. 2024) Furthermore, GIS helps identify areas with overlapping vulnerabilities, such as aged infrastructure located in flood-prone zones, enabling strategic planning for resilience and adaptation.

GIS for Emergency Response and Disaster Management

Infrastructure asset management is a critical component of urban and regional planning, involving the systematic process of maintaining, upgrading, and operating physical assets cost-effectively. (Hung V. V. et al. 2025) These assets include transportation networks, water supply systems, sewer lines, power grids, public buildings, parks, and communication systems—each of which plays a vital role in the quality of life and economic productivity of a community. Traditionally, asset management relied on disparate records, manual inspections, and limited geographic insights, often resulting in inefficient resource allocation, unanticipated system failures, and higher lifecycle costs. Geographic Information Systems (GIS) have fundamentally transformed this process by introducing spatial intelligence and comprehensive mapping capabilities that allow asset managers to visualize, monitor, and analyze infrastructure networks in real-time and with geographic precision. At its core, GIS is a framework for collecting, managing, and analyzing data that is linked to geographic locations. In the context of asset management, GIS allows for the precise mapping of infrastructure elements, associating them with detailed attribute data such as type, material, installation date, condition rating, maintenance history, and ownership. This spatially enabled asset inventory provides a centralized, visual interface for asset tracking and analysis, allowing infrastructure managers to see not just where assets are located but also how they perform, how they interact with their surroundings, and how they change over time. (Lawal O. O. et al. 2024) Through advanced geospatial analysis, GIS helps asset managers prioritize interventions, forecast deterioration, and support long-term investment planning.

Figure 1. Integration of BIM and GIS

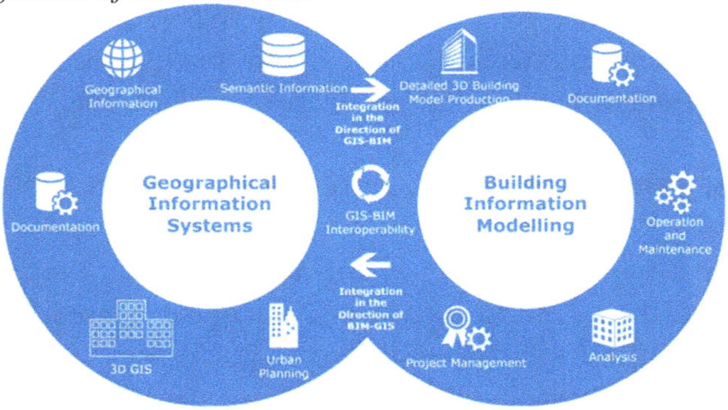

One of the most significant contributions of GIS to asset management is the creation and maintenance of a spatial asset registry. This registry maps every infrastructure asset onto a digital map, using standardized data layers that can be filtered, analyzed, and updated as needed. For example, a city's water utility can maintain a GIS-based map of its entire water distribution network, showing the exact location and attributes of pipelines, valves, hydrants, meters, and treatment facilities. (Hung V. V. et al. 2025) By integrating real-time sensor data and inspection results into the GIS platform, operators can detect leaks, identify pressure drops, and schedule targeted repairs—all with spatial awareness that improves efficiency and response times. In transportation infrastructure, GIS is used extensively to manage road networks, bridges, traffic signals, signage, and pavement conditions. Asset managers can use GIS tools to analyze pavement quality data, traffic volume, and environmental exposure to determine which roads require resurfacing or which bridges need structural assessments. These spatial insights help prioritize projects based on criticality, usage, and safety risk, ensuring that limited maintenance funds are allocated where they are most needed. Additionally, GIS facilitates the coordination of multiple infrastructure systems—such as synchronizing road repairs with utility upgrades—to minimize disruption and reduce costs. GIS also plays a vital role in risk assessment and resilience planning. By overlaying asset data with hazard maps—such as flood zones, seismic activity, storm surge areas, or wildfire risk zones—asset managers can assess the vulnerability of infrastructure assets to natural disasters. For instance, an electric utility can use GIS to identify which power lines and substations are located in high-risk wildfire areas and then plan preventive measures such as vegetation management, system hardening, or strategic shutoffs. (Lawal O. O. et al. 2024) This spatial risk analysis supports proactive planning, reduces downtime, and enhances public safety.

Integrating BIM and GIS: A Holistic Approach to Infrastructure Management

Geographic Information Systems (GIS) have emerged as indispensable tools in the realm of emergency response and disaster management, fundamentally transforming how governments, emergency services, humanitarian organizations, and communities prepare for, respond to, and recover from crises. (Misilmani A. H. et al. 2024) By providing a geospatial framework to collect, analyze, and visualize data in real time, GIS enables decision-makers to assess risks, allocate resources, coordinate responses, and communicate effectively during high-pressure situations. As the frequency and intensity of natural and human-made disasters continue to rise—driven by factors such as climate change, urbanization, and geopolitical instability—the ability to manage and mitigate emergencies through spatial intelligence is more critical than ever. (Li S. et al. 2024) At the core of GIS's application in emergency management is situational awareness. During a disaster, the speed and accuracy of decisions can be the difference between containment and catastrophe. GIS provides dynamic, interactive maps that synthesize vast volumes of information, including topography, population density, infrastructure networks, hazard zones, weather patterns, and the real-time location of emergency assets. (Li S. et al. 2024) This geospatial visualization allows emergency managers to quickly understand the scope of an incident, identify the most affected areas, and determine where to deploy personnel, equipment, and aid. For instance, during a wildfire, GIS can overlay satellite imagery with wind direction and terrain data to predict fire spread, guiding evacuation orders and resource deployment.

Pre-disaster planning is one of the most strategic uses of GIS in emergency management. Before any event occurs, GIS is used to map hazard-prone areas—such as floodplains, earthquake fault lines, landslide zones, or hurricane impact corridors—by analyzing historical data, environmental variables, and land use patterns. (Misilmani A. H. et al. 2024) These risk maps inform zoning regulations, infrastructure development, emergency preparedness plans, and insurance models. Urban planners and emergency agencies use GIS to conduct vulnerability assessments, determining which populations, facilities, and infrastructure are most at risk. (Wang X. et al. 2024) For example, a city might use GIS to identify elderly populations living in flood-prone areas who may require special assistance during an evacuation. These insights lead to more targeted and inclusive preparedness strategies. Emergency response coordination benefits immensely from GIS integration. During a disaster, multiple agencies—fire departments, police, medical responders, utility providers, and humanitarian organizations—must collaborate across jurisdictions and share accurate information. GIS platforms serve as centralized command systems where real-time data from various sources is consolidated and displayed. Incident command centers use GIS dashboards to monitor emergency calls, track responder locations, manage road closures, and assess damage reports. (Wang X. et al. 2024) The spatial context provided by GIS ensures that response teams are sent to the right places at the right times, reducing duplication of efforts and ensuring comprehensive coverage.

Benefits of BIM-GIS Integration: Improved Efficiency and Decision-Making

The integration of Building Information Modeling (BIM) and Geographic Information Systems (GIS) represents a transformative shift in infrastructure management, offering a holistic, data-driven framework that bridges the gap between detailed building-level information and broader geospatial context. (Matniyazov Z. et al. 2025) While BIM provides a rich, precise digital representation of the physical and functional characteristics of buildings and infrastructure assets, GIS offers a spatially comprehensive view of the environment in which these assets exist. Integrating these two technologies unifies micro-level and macro-level insights, enabling infrastructure stakeholders to make smarter, more coordinated, and context-aware decisions throughout the asset lifecycle—from planning and design to construction, operation, maintenance, and decommissioning. Traditionally, BIM and GIS have operated in separate domains. BIM has been widely adopted by architects, engineers, and construction professionals for the design and modeling of structures, delivering highly detailed 3D models with embedded data about materials, systems, and performance specifications. On the other hand, GIS has served urban planners, geographers, and government agencies, offering the tools to analyze terrain, manage networks, assess environmental impacts, and plan cities at a macro scale. The divergence of file formats, data structures, coordinate systems, and software platforms historically hindered effective interoperability between BIM and GIS. (Petojević Z. et al. 2025) However, with the growing complexity of infrastructure projects, climate challenges, urbanization pressures, and the demand for smart city solutions, the need for integration has become more urgent and achievable due to advancing technology and open data standards.

Integrating BIM and GIS allows stakeholders to view assets not only as isolated entities but as components within dynamic urban and ecological systems. For instance, a highway bridge modeled in BIM contains precise information about its structural components, materials, sensor networks, and maintenance history. (Matniyazov Z. et al. 2025) When integrated with GIS, that same bridge can be analyzed in terms of its connectivity to the regional road network, exposure to environmental hazards like floods or earthquakes, proximity to population centers, and role in evacuation routes. This added spatial intelligence supports resilience planning, asset prioritization, and lifecycle cost optimization. One of the key benefits of BIM-GIS integration is enhanced infrastructure planning and design. Urban planners and engineers can evaluate how proposed developments fit into existing environments, considering both the built and natural surroundings. For example, a hospital complex designed in BIM can be geolocated in GIS to assess its accessibility via public transport, its elevation relative to floodplains, or its adjacency to utilities and emergency services. Environmental impact assessments, zoning compliance checks, and traffic flow simulations become more comprehensive when the detail of BIM is contextualized within the geographic breadth of GIS. In construction and project management, integration improves coordination among stakeholders by linking construction schedules, material logistics, and site constraints with geographic data. GIS helps identify optimal access routes for transporting materials, maps construction impacts on nearby communities, and supports real-time tracking of equipment and personnel. (Petojević Z. et al. 2025) At the same time, BIM provides the precision needed for modeling structural assemblies, estimating quantities, and ensuring compliance with engineering specifications. A GIS-linked BIM model facilitates 4D (time) and 5D (cost) simulations that are grounded in real-world geographic conditions, leading to more predictable project outcomes and reduced risks.

Challenges and Opportunities: Implementing BIM-GIS Integration

The integration of Building Information Modeling (BIM) and Geographic Information Systems (GIS) heralds a new era of enhanced efficiency and superior decision-making in infrastructure development, urban planning, facility management, and environmental stewardship. Individually, BIM and GIS serve distinct but complementary functions—BIM provides highly detailed, data-rich digital representations of buildings and infrastructure assets, while GIS offers a broad spatial framework that maps these assets within their environmental, social, and economic contexts. (Carrasco C. A. et al. 2024) The seamless integration of these systems results in synergistic benefits that streamline workflows, optimize resource allocation, enable proactive management, and empower stakeholders with actionable intelligence. (Haidar C. H. et al. 2024) This fusion supports a more holistic approach to infrastructure lifecycle management, from initial design through construction to operations and maintenance, thereby significantly improving the efficiency and quality of decision-making processes. One of the primary benefits of BIM-GIS integration is improved data interoperability and visualization, which mitigates the long-standing challenges caused by siloed information systems. In many organizations, BIM data exists in isolated models focusing on architectural, structural, or mechanical components, while GIS databases maintain spatial data layers about land use, transportation networks, environmental features, and demographic patterns.

When these datasets are integrated, stakeholders gain access to a unified, spatially coherent repository that merges the micro-level precision of BIM with the macro-level context of GIS. This integration enhances visualization capabilities, allowing complex infrastructure projects to be viewed and analyzed in three dimensions and within their actual geographic environments. (Haidar C. H. et al. 2024) Decision-makers can explore interactive models that combine building specifications with site topography, utility layouts, and regional hazard maps, facilitating more informed and transparent choices.

Enhanced collaboration and communication constitute another major advantage of BIM-GIS integration. (Carrasco C. A. et al. 2024) Infrastructure projects typically involve multiple disciplines—architects, engineers, urban planners, environmental scientists, contractors, and facility managers—each relying on different data sources and tools. (Carrasco C. A. et al. 2024) Integrating BIM and GIS creates a common platform where all stakeholders can share consistent, up-to-date information, reducing miscommunications, data redundancies, and conflicts. (Carrasco C. A. et al. 2024) For example, during the planning phase of a new transportation corridor, engineers can use BIM to design bridges and tunnels with exact specifications, while GIS provides insights into land acquisition, environmental constraints, and connectivity to existing infrastructure. This comprehensive view fosters cross-disciplinary collaboration, accelerates consensus-building, and supports integrated planning efforts that consider technical feasibility, regulatory compliance, and community impact simultaneously. In terms of operational efficiency, BIM-GIS integration significantly improves asset management and maintenance processes. Traditional asset management systems often struggle to link detailed asset data with spatial location, leading to inefficiencies in tracking, inspection, and repair activities. (Ülger M. et al. 2023) By combining BIM's detailed asset attributes with GIS's spatial analytics, facility and infrastructure managers gain precise, location-aware digital twins—dynamic models that mirror real-world assets enriched with sensor data, inspection records, and performance metrics. This spatially enabled digital twin environment supports predictive maintenance by identifying deterioration patterns in relation to environmental factors such as soil conditions, weather exposure, or proximity to traffic loads. (Ülger M. et al. 2023) Maintenance scheduling becomes more efficient, reducing downtime and repair costs, while also extending asset lifespan and improving service reliability.

Applications of BIM and GIS in Urban Infrastructure

The integration of Building Information Modeling (BIM) and Geographic Information Systems (GIS) presents a transformative opportunity to revolutionize infrastructure management, urban planning, and facility operations by enabling a comprehensive and context-aware digital representation of built and natural environments. (ElBeheiry D. M. 2025) However, despite its vast potential, implementing BIM-GIS integration is a complex endeavor fraught with technical, organizational, and procedural challenges. At the same time, it opens significant opportunities for innovation, efficiency, sustainability, and stakeholder collaboration. Understanding these challenges and opportunities is critical for practitioners, policymakers, and technology developers striving to harness the full benefits of this powerful integration. One of the foremost challenges in BIM-GIS integration lies in the technical interoperability between these traditionally distinct systems. BIM models are highly detailed, object-oriented, and typically focused on vertical construction elements and asset-level information, often using proprietary or specialized file formats like IFC (Industry Foundation Classes), Revit (RVT), or Navisworks files. In contrast, GIS emphasizes geospatial data organized in layers, coordinate systems, and often less granular attribute data, stored in formats such as shapefiles, GeoJSON, or Esri File Geodatabases. (Hou Y. et al. 2023) Bridging these fundamentally different data structures, semantics, and spatial referencing schemes requires complex data transformation, alignment, and harmonization processes. For example, reconciling BIM's local coordinate systems with GIS's global geographic coordinate frameworks demands precise georeferencing techniques and standards. (Hou Y. et al. 2023) Without robust interoperability protocols and middleware solutions, integrating BIM and GIS data risks inconsistencies, data loss, or reduced fidelity, which can undermine decision-making accuracy.

Another significant technical hurdle is the data volume and complexity involved. BIM models can contain millions of detailed components and metadata, creating massive datasets that challenge storage, processing, and real-time rendering capabilities, especially when combined with extensive GIS layers covering large geographic areas. Managing this "big data" efficiently requires advanced data management architectures, cloud computing, and high-performance visualization tools. (ElBeheiry D. M. 2025) Furthermore, integrating dynamic data sources—such as real-time sensor feeds, IoT devices, or remote sensing imagery—with static BIM and GIS datasets introduces additional layers of complexity, necessitating scalable platforms that can handle streaming data and temporal analyses. Standardization and data quality issues also pose critical challenges. (Saksena H. T. 2024) Although open standards like IFC for BIM and OGC standards for GIS exist, variations in adoption, inconsistent implementation, and gaps in standard coverage often hinder seamless integration. Inconsistent attribute schemas, naming conventions, and data completeness between BIM and GIS datasets can lead to semantic mismatches and interoperability breakdowns. Ensuring high data quality, currency, and accuracy across both BIM and GIS requires rigorous data governance frameworks and coordinated workflows among diverse stakeholders. (Saksena H. T. 2024) Without effective quality control mechanisms, integrated datasets may propagate errors, impacting downstream applications such as facility management, emergency response, or environmental modeling.

Case Studies: Successful Implementation of BIM and GIS in Urban Infrastructure

The integration and application of Building Information Modeling (BIM) and Geographic Information Systems (GIS) in urban infrastructure represent a powerful confluence of detailed asset modeling and expansive spatial analysis, which are crucial to managing and optimizing key sectors such as transportation, water, and energy. (Kaewpoonsuk P. et al. 2024) These sectors form the backbone of modern cities and are increasingly challenged by rapid urbanization, climate change, aging infrastructure, and evolving regulatory requirements. The (Andritsou D. et al. 2024) combined use of BIM and GIS technologies not only enhances the design, construction, and operation of infrastructure but also enables more intelligent, sustainable, and resilient urban systems that improve quality of life for citizens while reducing costs and environmental impacts. In the transportation sector, BIM and GIS play complementary roles that transform how roadways, railways, airports, transit systems, and pedestrian networks are planned, designed, constructed, and maintained. (Andritsou D. et al. 2024) BIM provides precise 3D models of infrastructure components such as bridges, tunnels, stations, and track systems, including detailed specifications of materials, structural elements, utilities, and mechanical systems. GIS situates these components within a larger geographic context, integrating topography, land use, traffic patterns, population density, environmental constraints, and other spatial data. For example, during the planning phase of a new highway project, GIS tools analyze terrain, existing road networks, environmental impact zones, and demographic data to identify optimal routes that minimize social and ecological disruptions. BIM models then refine structural designs, simulate construction sequences, and enable clash detection and cost estimation. (Barrile V. et al. 2024) During construction and operation, GIS supports real-time monitoring of traffic flows, congestion hotspots, and incident management, while BIM aids in maintenance scheduling, asset condition assessments, and lifecycle management. The integration of BIM-GIS data facilitates multimodal transportation planning by combining detailed asset information with urban mobility patterns, allowing cities to optimize public transit routes, reduce emissions, and enhance accessibility.

The water infrastructure sector—encompassing supply networks, wastewater systems, stormwater management, and treatment facilities—also benefits profoundly from BIM and GIS integration. (Kaewpoonsuk P. et al. 2024) GIS provides the spatial framework for mapping water sources, pipelines, pump stations, reservoirs, and drainage basins, overlaying environmental data such as rainfall patterns, flood zones, soil types, and watershed boundaries. This geospatial insight is essential for identifying vulnerabilities, planning expansions, and managing resources sustainably. (Barrile V. et al. 2024) BIM contributes detailed, component-level models of water treatment plants, pumping equipment, valves, and pipes, including metadata on materials, capacities, maintenance history, and operational parameters. For instance, in a municipal water distribution system, GIS can identify areas prone to pressure drops or contamination risks by analyzing spatial patterns of demand, topography, and pipe networks, while BIM models allow engineers to simulate hydraulic performance and detect potential system failures. During emergency events such as floods or pipeline bursts, integrated BIM-GIS platforms enable rapid damage assessment, repair prioritization, and coordination of response efforts. Moreover, the fusion of BIM and GIS supports predictive maintenance by correlating spatial environmental data with asset conditions, allowing utilities to schedule interventions before failures occur, thereby reducing water loss, service interruptions, and repair costs. Smart water management also leverages sensor data integrated into BIM-GIS digital twins to monitor water quality, detect leaks, and optimize energy use for pumping and treatment.

Best Practices: Implementing BIM and GIS in Urban Infrastructure Management

The practical integration of Building Information Modeling (BIM) and Geographic Information Systems (GIS) in urban infrastructure projects has transitioned from theoretical potential to tangible reality, demonstrated through a growing number of successful case studies worldwide. (Demir S. et al. 2024) These real-world implementations showcase the ability of BIM-GIS integration to enhance project delivery, operational efficiency, stakeholder collaboration, and urban resilience across diverse infrastructure sectors such as transportation, water management, energy systems, and public safety. (Piras G. et al. 2024) Examining these cases reveals best practices, lessons learned, and emerging trends that guide future applications and promote wider adoption of integrated digital technologies in urban infrastructure management. One prominent example is the redevelopment of the city of Helsinki's Railway Station Area project in Finland. This ambitious urban renewal initiative integrated BIM and GIS to facilitate a complex mix of infrastructure upgrades, new commercial and residential developments, and public space improvements. Helsinki's city planners and engineers used GIS to analyze the spatial relationships between existing transportation networks, underground utilities, and land use patterns, while BIM was employed to develop detailed 3D models of proposed buildings, tunnels, and station facilities. The combined use of these technologies enabled comprehensive clash detection, optimized construction sequencing, and facilitated stakeholder engagement through interactive 3D visualizations accessible to the public. By integrating spatial and asset data, the project minimized disruptions during construction, improved coordination among contractors and authorities, and achieved significant cost savings. This case exemplifies how BIM-GIS integration supports urban densification while preserving functionality and accessibility in highly complex environments.

In the realm of water infrastructure, the city of Singapore has set a benchmark with its use of integrated BIM and GIS in managing its extensive water supply and drainage networks. (Ismaeil E. M. 2024) Singapore's Public Utilities Board (PUB) employs a sophisticated digital twin platform that combines detailed BIM models of water treatment plants, pipelines, valves, and pumping stations with GIS-based spatial data on topography, land use, and hydrological patterns. This integration facilitates real-time monitoring of water (Demir S. et al. 2024) quality, leak detection, and flood risk management. During heavy rainfall events, the system allows rapid identification of vulnerable drainage areas and prioritizes maintenance interventions to prevent urban flooding. Moreover, PUB uses BIM-GIS integration to simulate future demand scenarios based on urban growth projections, optimizing the expansion of water infrastructure. Singapore's success highlights the value of integrated digital models in enhancing urban resilience and sustainability in a water-scarce context. The energy sector provides another illustrative case from the city of Amsterdam in the Netherlands, where the municipality and utility companies have collaborated on a smart grid initiative leveraging BIM and GIS integration. Amsterdam's integrated platform maps energy generation assets—including solar panels, wind turbines, and substations—and overlays them with GIS data on energy demand distribution, population density, and urban microclimates. (Di Giuda G. M. et al. 2024) BIM models provide detailed technical specifications of energy equipment and infrastructure condition data. The combined dataset supports dynamic load balancing, outage prediction, and targeted maintenance. (Di Giuda G. M. et al. 2024) For example, during peak demand periods, operators use the platform to identify distributed energy resources that can be activated to reduce grid strain, improving energy efficiency and reducing carbon emissions. Amsterdam's project demonstrates how BIM-GIS integration can advance smart city goals by creating adaptive, low-carbon energy systems tailored to urban spatial and temporal dynamics.

The Future of Urban Infrastructure Management: Leveraging BIM and GIS

The integration of Building Information Modeling (BIM) and Geographic Information Systems (GIS) marks a significant paradigm shift in urban infrastructure management, offering unprecedented opportunities to enhance the planning, design, construction, operation, and maintenance of complex urban systems. As cities worldwide grapple with increasing urbanization, aging infrastructure, environmental challenges, and rising demands for sustainable and resilient services, the synergy between BIM and GIS emerges as a critical enabler for smarter, more efficient, and adaptive urban infrastructure management. This conclusion synthesizes the key insights gained from exploring the role, applications, best practices, and case studies related to BIM-GIS integration, underscoring its transformative potential and future directions. At the core, BIM provides detailed, asset-level digital representations of infrastructure components, capturing their geometry, spatial relationships, material properties, and lifecycle information. GIS, on the other hand, offers powerful tools for spatial analysis, mapping, and contextualizing these assets within their geographic, environmental, and socio-economic settings. The fusion of BIM's precision with GIS's spatial intelligence creates comprehensive digital twins of urban infrastructure that reflect both the physical and contextual realities of complex urban systems. This integration enables stakeholders to visualize, analyze, and manage infrastructure assets more holistically, enhancing decision-making throughout the entire asset lifecycle—from conceptual design and construction to operation, maintenance, and eventual decommissioning. One of the most profound potentials of BIM-GIS integration lies in its ability to support data-driven, evidence-based decision-making. By combining detailed BIM data with GIS spatial analytics, urban planners, engineers, and facility managers gain deeper insights into infrastructure performance, vulnerability, and interactions within the urban environment.

This facilitates optimized resource allocation, risk mitigation, and proactive maintenance strategies that reduce operational costs, extend asset lifespans, and improve service reliability. Moreover, integrated models support scenario planning and simulation for future urban growth, climate change adaptation, emergency response, and sustainability initiatives. Such forward-looking capabilities enable cities to transition from reactive to predictive and adaptive infrastructure management approaches. The application domains of BIM and GIS integration span critical urban infrastructure sectors including transportation, water, and energy systems, each benefiting uniquely from this digital synergy. Transportation projects leverage integrated BIM-GIS to optimize route planning, construction coordination, and traffic management, enhancing mobility and reducing congestion. Water utilities employ these technologies for detailed asset tracking, hydraulic modeling, and flood risk management, thereby improving water security and resilience. Energy systems utilize integrated models for smart grid management, renewable energy integration, and demand forecasting, supporting the transition to low-carbon urban energy systems. Beyond sectoral benefits, BIM-GIS integration fosters cross-sectoral coordination, addressing the interdependencies among infrastructure networks and promoting more coherent urban development and emergency preparedness. Successful case studies from global cities demonstrate the tangible benefits and feasibility of BIM-GIS integration, highlighting how these technologies can overcome traditional data silos, enhance stakeholder collaboration, and improve project outcomes.

CONCLUSION

Implementing Building Information Modeling (BIM) and Geographic Information Systems (GIS) in urban infrastructure management presents a transformative opportunity to enhance planning, design, construction, operation, and maintenance of urban assets. However, to realize the full potential of BIM-GIS integration, it is essential to adopt a set of best practices that address technical, organizational, and procedural complexities inherent in combining these technologies. These best practices, distilled from extensive research and real-world applications, serve as guidelines to ensure interoperability, data quality, stakeholder collaboration, and sustainable workflows that maximize benefits while minimizing risks and inefficiencies. The foundation of successful BIM-GIS implementation lies in establishing clear data interoperability standards and protocols. Given the fundamentally different data models, formats, and coordinate systems used by BIM and GIS platforms, it is critical to develop and adhere to open, standardized data exchange frameworks. Industry standards such as IFC (Industry Foundation Classes) for BIM and OGC (Open Geospatial Consortium) standards for GIS provide a starting point, but customized mappings and middleware solutions are often necessary to bridge semantic and structural gaps. Early adoption of these standards during project inception ensures seamless data integration, reduces errors and redundancy, and facilitates automated workflows. Additionally, ongoing collaboration with standardization bodies and software vendors helps keep integration capabilities current with evolving technologies. Comprehensive and rigorous data governance is another pillar of best practice, encompassing data quality assurance, version control, access management, and metadata documentation. High data quality—accurate, complete, consistent, and up-to-date—is paramount for reliable analysis and decision-making. Establishing protocols for data validation, error checking, and update frequency prevents the propagation of inaccuracies across integrated BIM-GIS datasets. Version control mechanisms track changes to models and spatial data, enabling traceability and rollback if necessary. Clear policies defining user roles, permissions, and data security protect sensitive information while promoting appropriate data sharing among stakeholders. Metadata standards enhance data discoverability and contextual understanding, supporting effective reuse and integration. Early and sustained stakeholder engagement and interdisciplinary collaboration are essential for overcoming organizational silos and aligning objectives across diverse teams involved in urban infrastructure projects. BIM traditionally serves architects, engineers, and contractors, while GIS is often managed by planners, environmental analysts, and municipal authorities. Bridging these cultural and functional divides requires structured communication channels, joint workshops, and shared digital platforms that foster transparency and mutual understanding. Engaging end-users such as facility managers, maintenance crews, and community representatives early in the process ensures that integrated BIM-GIS solutions address practical needs and enhance usability. Collaborative governance models and integrated project delivery (IPD) frameworks that incentivize cooperation contribute significantly to project success.

REFERENCES

Akindele, O., Ajayi, S., Oyegoke, A. S., Alaka, H. A., & Omotayo, T. (2025). Application of geographic information system (GIS) in construction: A systematic review. *Smart and Sustainable Built Environment*, *14*(1), 210–236. DOI: 10.1108/SASBE-01-2023-0016

Andritsou, D., & Potsiou, C. (2024). BIM-GIS integration for interactive, open, and low-cost 3D land use registration and urban neighborhood management. *Tehnički glasnik*, 18(si1), 67-74.

Barrile, V., La Foresta, F., Calcagno, S., & Genovese, E. (2024). Innovative system for BIM/GIS integration in the context of urban sustainability. *Applied Sciences (Basel, Switzerland)*, *14*(19), 8704. DOI: 10.3390/app14198704

Borkowski, A. S., & Olszewska, P. (2025). BIM model of district heating networks in design and investment management processes: A case study. *Sustainability (Basel)*, *17*(9), 4102. DOI: 10.3390/su17094102

Carrasco, C. A., Lombillo, I., Sánchez-Espeso, J. M., & Balbás, F. J. (2024). A new approach to 3D facilities management in buildings using GIS and BIM integration: A case study application. *Applied Sciences (Basel, Switzerland)*, *14*(23), 11097. DOI: 10.3390/app142311097

Carrasco, C. A., Lombillo, I., Sánchez-Espeso, J. M., Blanco, H., & Boffill, Y. (2024). Methodology for 3D management of university faculties using integrated GIS and BIM models: A case study. *Buildings (Basel, Switzerland)*, *14*(11), 3547. DOI: 10.3390/buildings14113547

Cepa, J. J., Alberti, M. G., Pavón, R. M., & Calvo, J. A. (2024). Integrating BIM and GIS for an existing infrastructure. *Applied Sciences (Basel, Switzerland)*, *14*(23), 10962. DOI: 10.3390/app142310962

Chib, S., Mehta, A. K., Selvakumar, P., Saravanan, A., Mishra, B. R., & Manjunath, T. C. (2025). Mental health challenges and solutions in high-pressure work environments. In Harriott, S. (Ed.), *Innovative Approaches to Managing Conflict and Change in Diverse Work Environments* (pp. 247–272). IGI Global Scientific Publishing. DOI: 10.4018/979-8-3693-9556-1.ch011

Dahiya, R., & Selvakumar, P. Vidyadhara, & Manjunath, T. C. (2025). Design and aerodynamics: Principles and advances. In M. Mellal (Ed.), *Innovations and Developments in Unmanned Aerial Vehicles* (pp. 145-170). IGI Global Scientific Publishing.

Demir, S., & Yomralioglu, T. (2024). Bridging geo-data and natural gas pipeline design standards: A systematic review of BIM-GIS integration for natural gas pipeline asset management. *Energies*, *17*(10), 2306. DOI: 10.3390/en17102306

Di Giuda, G. M., Meschini, S., Gasbarri, P., Accardo, D., Tagliabue, L. C., & Cacciaguerra, E. (2024). BIM, GIS and BI tools for a university asset management system supporting space management, occupancy evaluation and optimization strategies. [ITcon]. *Information Technology and Control*, *29*, 1128–1155.

ElBeheiry, D. M. (2025). An integrated BIM and GIS framework for sustainable urban planning. *JES. Journal of Engineering Sciences*, *53*(3), 270–287.

Gao, Y., Xiong, G., Hu, Z., Chai, C., & Li, H. (2024). Bridge digital twin for practical bridge operation and maintenance by integrating GIS and BIM. *Buildings (Basel, Switzerland)*, *14*(12), 3731. DOI: 10.3390/buildings14123731

Haidar, C. H., Maqsoom, A., Choudhary, S., Shahzad, M., & Khalid, M. H. (2024, December). Rainwater harvesting for urban flood mitigation: A BIM-GIS integrated approach. In *2024 International Conference on Frontiers of Information Technology (FIT)* (pp. 1-6). IEEE. DOI: 10.1109/FIT63703.2024.10838457

Hou, Y., Heo, J., & Chang, S. (2023). Information visualization platform for underground pipeline maintenance integrating BIM and GIS data. In *Computing in Civil Engineering 2023* (pp. 781-789).

Hung, V. V., & Phuong, N. C. (2025). Integration of BIM and GIS on web viewer environment for Tan Cang station asset management. *Transportation Research Procedia*, *85*, 281–288. DOI: 10.1016/j.trpro.2025.03.163

Ismaeil, E. M. (2024). Asset information model management-based GIS/BIM integration in facility management contract. *Sustainability (Basel)*, *16*(6), 2495. DOI: 10.3390/su16062495

Kaewpoonsuk, P., & Subsomboo, K. (2024). Urban information modeling for smart cities: Advancing sustainable infrastructure and environmental management in digital twin applications. *Pakistan Journal of Life and Social Sciences*, *22*(2). Advance online publication. DOI: 10.57239/PJLSS-2024-22.2.001545

Kaur, M., Selvakumar, P., & Dinesh, N. Mohit, Chandel, P. S., & Manjunath, T. C. (2025). Reconstruction and preservation strategies for cultural heritage. In *S. Oflazoğlu & M. Dora (Eds.), Preserving Cultural Heritage in Post-Disaster Urban Renewal* (pp. 385-412). IGI Global Scientific Publishing.

Kommineni, M., Chundru, S., Maroju, P. K., & Selvakumar, P. (2025). Ethical implications of AI in sustainable development pedagogy. In Chinedu, C., Mafarja, N., Hussein, M., & Saleem, A. (Eds.), *Rethinking the Pedagogy of Sustainable Development in the AI Era* (pp. 17–36). IGI Global Scientific Publishing. DOI: 10.4018/979-8-3693-9062-7.ch002

Lawal, O. O., & Nawari, N. O. (2024). Blockchain-enabled city information modeling framework for urban asset management. *Journal of Architectural Engineering*, *30*(3), 04024020. DOI: 10.1061/JAEIED.AEENG-1644

Lee, J., Lee, J. H., Im, J., & Jang, J. (2024). Enhancing urban safety: Optimal patrol route strategies for volunteer security squads based on integrated BIM-GIS data. *Buildings (Basel, Switzerland)*, *14*(10), 3074. DOI: 10.3390/buildings14103074

Li, S., Suh, W., & Kim, J. I. (2024). An integrated urban road rehabilitation planning and decision support based on BIM and GIS. In *Computing in Civil Engineering 2023* (pp. 112-120). DOI: 10.1061/9780784485231.014

Limbore, N. V., Anute, N., Ahluwalia, G. K., Chordiya, P. B., Kalshetti, P., & Selvakumar, P. (2025). Ethical leadership and corporate responsibility. In Harriott, S. (Ed.), *Innovative Approaches to Managing Conflict and Change in Diverse Work Environments* (pp. 193–218). IGI Global Scientific Publishing. DOI: 10.4018/979-8-3693-9556-1.ch009

Matniyazov, Z., Adilov, Z., Khotamov, A., Elmurodov, S., Rasul-Zade, L., & Abdikhalilov, F. (2025). Integration of BIM and GIS technologies in modern urban planning: Challenges and prospects. *American Journal of Education and Learning*, *3*(3), 972–977.

Misilmani, A. H., El-Bastawissi, I. Y., Ayad, H., & El Sayary, S. (2024). BIM-GIS integration: An innovative tool to enhance urban heritage management in the digital era. [APJ]. *Architecture and Planning Journal*, *30*(1), 3. DOI: 10.54729/2789-8547.1245

Mushtaha, A. W., Alaloul, W. S., Musarat, M. A., Baarimah, A. O., Rabah, F. K., & Alawag, A. M. (2024, January). BIM-GIS integration for infrastructure management in post-disaster stage. In *2024 ASU International Conference in Emerging Technologies for Sustainability and Intelligent Systems (ICETSIS)* (pp. 856-861). IEEE. DOI: 10.1109/ICETSIS61505.2024.10459527

Pelden, S., Banihashemi, S., Mohandes, S. R., Arashpour, M., & Kalantari, M. (2025). Enhancing infrastructure planning and design through BIM-GIS integration. *Structure and Infrastructure Engineering*, •••, 1–20. DOI: 10.1080/15732479.2025.2461629

Petojević, Z., Parezanović, A., Nadaždi, A., Isailović, D., & Višnjevac, N. (2025). Integration of BIM and GIS to reduce construction and demolition waste. In *Advances in Construction and Demolition Waste Recycling* (pp. 99–118). Woodhead Publishing. DOI: 10.1016/B978-0-443-23962-5.00002-5

Piras, G., Agostinelli, S., & Muzi, F. (2024). Digital twin framework for built environment: A review of key enablers. *Energies*, *17*(2), 436. DOI: 10.3390/en17020436

Piras, G., Muzi, F., & Zylka, C. (2024). Integration of BIM and GIS for the digitization of the built environment. *Applied Sciences (Basel, Switzerland)*, *14*(23), 11171. DOI: 10.3390/app142311171

Popov, P., Mozaffari, M. H., RazaviAlavi, S. R., & Jalaei, F. (2025). GIS-based solutions for management of public building and infrastructure assets: A review of state of the art and research trend analysis. *Smart Construction*, *2*(2), 1–2. DOI: 10.55092/sc20250009

Rotilio, M., Marchionni, C., Massari, A., Di Giovanni, G., & Simeone, D. (2024, June). Efficiency management of built environment: The digitalization of energy and service grids. In *2024 IEEE International Workshop on Metrology for Living Environment (MetroLivEnv)* (pp. 355-360). IEEE. DOI: 10.1109/MetroLivEnv60384.2024.10615404

Saksena, H. T. (2024). Enhancing Indonesia's transportation infrastructure planning through BIM and digital twins. *FSTPT International Journal of Transportation Studies*, *1*(2), 36–48.

Samal, A., & Misra, L. I. Mohit, Pougajendy, S., Chandel, P. S., Manjunath, T. C., & Selvakumar, P. (2025). Musealization and tourism practices. In *S. Oflazoğlu & M. Dora (Eds.), Preserving Cultural Heritage in Post-Disaster Urban Renewal* (pp. 115-142). IGI Global Scientific Publishing.

Sivakumar, A., Melina, J. V. T., & Suresh, E. S. M. (2025). BIM-GIS integration for efficient management of urban underground stormwater networks. *Proceedings of the Institution of Civil Engineers-Smart Infrastructure and Construction*, 1-14. DOI: 10.1680/jsmic.23.00092

Sohaib, M., Najeeb, A., Umair, M., Khan, M. A., Zubair, M. U., Jehan, Z., & Khattak, A. (2024). Improving urban road infrastructure analysis and design using an integrated BIM-GIS and traffic microsimulation framework. *Innovative Infrastructure Solutions*, *9*(7), 285. DOI: 10.1007/s41062-024-01609-z

Soundariya, A., Behera, N. K., Selvakumar, P., Chandel, P. S., Gupta, S., & Manjunath, T. C. (2025). AI and environmental ethics: The role of AI in addressing climate change ethical considerations in AI-driven environmental solutions. In Chakraborty, S. (Ed.), *Human Values, Ethics, and Dignity in the Age of Artificial Intelligence* (pp. 321–346). IGI Global Scientific Publishing. DOI: 10.4018/979-8-3693-9035-1.ch013

Sreekanthaswamy, N., Anitha, S., Singh, A., Jayadeva, S. M., Gupta, S., Manjunath, T. C., & Selvakumar, P. (2025). Digital tools and methods. In Çela, E., Paliktzoglou, V., Sengar, S., Eappen, P., & Vajjhala, N. (Eds.), *Enhancing School Counseling With Technology and Case Studies* (pp. 25–48). IGI Global Scientific Publishing. DOI: 10.4018/979-8-3693-8392-6.ch002

Tamilarasi, C., Sowmiya, K., Vignesh, G. D., Jasmine, R. L., Nehru, P., & Selvakumar, P. (2025). AI in renewable energy education: Teaching the future of sustainable power systems. In Chinedu, C., Mafarja, N., Hussein, M., & Saleem, A. (Eds.), *Rethinking the Pedagogy of Sustainable Development in the AI Era* (pp. 133–152). IGI Global Scientific Publishing. DOI: 10.4018/979-8-3693-9062-7.ch007

Ülger, M., & Kocaman, H. (2023). A new modeling approach with integrated use of BIM and GIS in rural resettlement areas. Available at *SSRN* 5006381.

Van Tan, N., Le Vin, T., Van Do, T., & Thao, N. D. (2025, May). Integration of advanced surveying technologies into BIM-GIS systems of port infrastructure. In *Proceedings of 7th International Conference on Civil Engineering and Architecture,* Volume 1*: Proceedings of ICCEA 2024* (Vol. 640, p. 178). Springer Nature.

Wang, X., & Xie, M. (2024). Integration of 3DGIS and BIM and its application in visual detection of concealed facilities. *Geo-Spatial Information Science*, *27*(1), 132–141. DOI: 10.1080/10095020.2022.2054732

Compilation of References

Aboulnaga, M., Abouaiana, A., Puma, P., Elsharkawy, M., Farid, M., Gamal, S., & Lucchi, E. (2024). Climate change and cultural heritage: A global mapping of the UNESCO thematic indicators in conjunction with advanced technologies for cultural sustainability. *Sustainability (Basel)*, *16*(11), 4650. DOI: 10.3390/su16114650

Adeyemi, A. B., Ohakawa, T. C., Okwandu, A. C., Iwuanyanwu, O., & Ifechukwu, G. O. (2024). Advanced Building Information Modeling (BIM) for affordable housing projects: Enhancing design efficiency and cost management. *Journal of Building Information Modeling*, *12*(1), 45–60.

Ahmad, K., Maabreh, M., Ghaly, M., Khan, K., Qadir, J., & Al-Fuqaha, A. (2022). Developing future human-centered smart cities: Critical analysis of smart city security, Data management, and Ethical challenges. *Computer Science Review*, *43*, 100452. DOI: 10.1016/j.cosrev.2021.100452

Ahmad, M. (2023). AI-enabled spatial intelligence: Revolutionizing data management and decision making in geographic information systems. In *AI and Its Convergence With Communication Technologies* (pp. 137–166). IGI Global.

Ait-Lamallam, S., Sebari, I., Yaagoubi, R., & Doukari, O. (2019). Vers une extension du Building Information Modeling aux projets d'infrastructure routière: Cas d'étude d'un échangeur autoroutier au Maroc. *Revue Marocaine des Sciences Agronomiques et Vétérinaires*, *28*(1), 609–617.

Akindele, O., Ajayi, S., Oyegoke, A. S., Alaka, H. A., & Omotayo, T. (2025). Application of Geographic Information System (GIS) in construction: A systematic review. *Smart and Sustainable Built Environment*, *14*(1), 210–236. DOI: 10.1108/SASBE-01-2023-0016

Akob, Z., Hipni, M. Z. A., & Abd Razak, A. A. A. (2019). Deployment of GIS + BIM in the construction of Pan Borneo Highway Sarawak, Malaysia. [IOP Publishing.]. *IOP Conference Series. Materials Science and Engineering*, *512*, 012037. DOI: 10.1088/1757-899X/512/1/012037

Al Salmi, H., & Abdulmuttalib, H. M. (2012). ROLE ASSESSMENT OF GIS ANALYSIS AND ITS RELIABILITY WHILE RANKING URBAN SUSTAINABILITY USING SCENARIOS SPECIFIC TO REGIONAL CLIMATE, COMMUNITY AND CULTURE. *The International Archives of the Photogrammetry. Remote Sensing and Spatial Information Sciences, XXXIX, B4*, 321–326. DOI: 10.5194/isprsarchives-XXXIX-B4-321-2012

Alahacoon, N., & Edirisinghe, M. (2021). Spatial variability of rainfall trends in sri lanka from 1989 to 2019 as an indication of climate change. *ISPRS International Journal of Geo-Information*, *10*(2), 84. Advance online publication. DOI: 10.3390/ijgi10020084

Alberti, M., & Susskind, L. (1996). Managing urban sustainability: An introduction to the special issue. *Environmental Impact Assessment Review*, *16*(4–6), 213–221. DOI: 10.1016/S0195-9255(96)00070-4

Alharbi, A. A. (2025). Cognitive learning approach to enhance university students' visualization of molecular geometry in chemical compounds: A case study in Saudi Arabia. *Journal of Radiation Research and Applied Sciences*, *18*(1), 101283. DOI: 10.1016/j.jrras.2024.101283

Alhazmi, H., & Loy, A. C. M. (2021). A review on environmental assessment of conversion of agriculture waste to bio-energy via different thermochemical routes: Current and future trends. *Bioresource Technology Reports*, *14*, 100682. DOI: 10.1016/j.biteb.2021.100682

Ali, K. N., Alhajlah, H. H., & Kassem, M. A. (2022). Collaboration and Risk in Building Information Modelling (BIM): A Systematic Literature Review. *Buildings (Basel, Switzerland)*, *12*(5), 571. Advance online publication. DOI: 10.3390/buildings12050571

Alizadeh, K., Bajgiran, M., & Nik, E. (2016). Building Information Modeling (BIM): A study to prioritize applications, risks and challenges. *International Conference on Industrial Engineering and Operations Management*, (18), 1–7.

Aljoufie, M. (2013). Toward integrated land use and transport planning in fast-growing cities: The case of Jeddah, Saudi Arabia. *Habitat International*, *41*, 205–215. DOI: 10.1016/j.habitatint.2013.08.010

Alkaissi, Z. A. (2024). Traffic congestion evaluation of urban streets based on fuzzy inference system and GIS application. *Ain Shams Engineering Journal, 15*(6), 102725. DOI: 10.1016/j.asej.2024.102725

Al-Mhdawi, M. K. S., Dacre, N., Brito, M., Baxter, D., Xu, K., & Young, C. (2023, March). *An agile compliance framework for the European cooperation for space standardization. In 2023 IEEE Aerospace Conference*. IEEE.

Al-Shihri, F. S. (2013). PRINCIPLES OF SUSTAINABLE DEVELOPMENT AND THEIR APPLICATION IN URBAN PLANNING IN SAUDI ARABIA. *JES. Journal of Engineering Sciences, 41*(4), 1703–1727. DOI: 10.21608/jesaun.2013.114898

Alsubai, S., Dutta, A. K., & Sait, A. R. W. (2024). Hybrid deep learning-based traffic congestion control in IoT environment using enhanced arithmetic optimization technique. *Alexandria Engineering Journal, 105*, 331–340. DOI: 10.1016/j.aej.2024.06.072

Ameen, R. F. M., Mourshed, M., & Li, H. (2015). A critical review of environmental assessment tools for sustainable urban design. *Environmental Impact Assessment Review, 55*, 110–125. DOI: 10.1016/j.eiar.2015.07.006

Amirebrahimi, S., & Rajabifard, A. (2012). AN INTEGRATED WEB-BASED 3D MODELING AND VISUALIZATION PLATFORM TO SUPPORT SUSTAINABLE CITIES. *ISPRS Annals of the Photogrammetry, Remote Sensing and Spatial Information Sciences, I–4*, 299–304. DOI: 10.5194/isprsannals-I-4-299-2012

Anastasovitis, E., Georgiou, G., Matinopoulou, E., Nikolopoulos, S., Kompatsiaris, I., & Roumeliotis, M. (2024). Enhanced inclusion through advanced immersion in cultural heritage: A holistic framework in virtual museology. *Electronics (Basel), 13*(7), 1396. DOI: 10.3390/electronics13071396

Andrianesi, D., & Dimopoulou, E. (2020). An integrated bim-gis platform for representing and visualizing 3d cadastral data. Isprs Annals of the Photogrammetry Remote Sensing and Spatial Information Sciences, VI-4/W1-2020, 3-11. DOI: 10.5194/isprs-annals-VI-4-W1-2020-3-2020

Andritsou, D., & Potsiou, C. (2024). BIM-GIS integration for interactive, open, and low-cost 3D land use registration and urban neighborhood management. *Tehnički glasnik*, 18(si1), 67-74.

Arampatzis, G., Kiranoudis, C., Scaloubacas, P., & Assimacopoulos, D. (2003). A GIS-based decision support system for planning urban transportation policies. *European Journal of Operational Research, 152*(2), 465–475. DOI: 10.1016/S0377-2217(03)00037-7

Aslam, M. M., Shafik, W., Hidayatullah, A. F., Kalinaki, K., Gul, H., Zakari, R. Y., & Tufail, A. (2025). Intelligent Transportation Systems: A Critical review of integration of Cyber-physical Systems (CPS) and Industry 4.0. *Digital Communications and Networks*. Advance online publication. DOI: 10.1016/j.dcan.2025.06.014

Avezbaev, S., Avezbaev, O., Tashpulatov, S., & Sharipov, S. (2023). Implementation of gis-based smart community information system and concepts of digital twin in the field of urban planning in uzbekistan. E3s Web of Conferences, 386, 05006. DOI: 10.1051/e3sconf/202338605006

Azari, P., Li, S., Shaker, A., & Sattar, S. (2025). Georeferencing Building Information Models for BIM/GIS Integration: A Review of Methods and Tools. *ISPRS International Journal of Geo-Information*, *14*(5), 180. DOI: 10.3390/ijgi14050180

Babalola, A., Musa, S., Akinlolu, M. T., & Haupt, T. C. (2023). A bibliometric review of advances in building information modeling (BIM) research. *Journal of Engineering. Defence Technology*, *21*(3), 690–710. DOI: 10.1108/JEDT-01-2021-0013

Bachechi, C., Po, L., & Rollo, F. (2021). Big data analytics and visualization in traffic monitoring. *Big Data Research*, *27*, 100292. DOI: 10.1016/j.bdr.2021.100292

Banani, R., Vahdati, M. M., Shahrestani, M., & Clements-Croome, D. (2016). The development of building assessment criteria framework for sustainable non-residential buildings in Saudi Arabia. *Sustainable Cities and Society*, *26*, 289–305. DOI: 10.1016/j.scs.2016.07.007

Barrile, V., La Foresta, F., Calcagno, S., & Genovese, E. (2024). Innovative system for BIM/GIS integration in the context of urban sustainability. *Applied Sciences (Basel, Switzerland)*, *14*(19), 8704. DOI: 10.3390/app14198704

Basir, W., Ujang, U., & Majid, Z. (2023). Adaptation 4d and 5d bim for bim/gis data integration in construction project management. *IOP Conference Series. Earth and Environmental Science*, *1274*(1), 012002. DOI: 10.1088/1755-1315/1274/1/012002

Basu, A. J., & Kumar, U. (2004). Innovation and Technology Driven Sustainability Performance Management Framework (ITSPM) for the Mining and Minerals Sector. *International Journal of Surface Mining. Reclamation and Environment*, *18*(2), 135–149. DOI: 10.1080/13895260412331295394

Batty, M. (2018). Digital twins. *Environment and Planning. B, Urban Analytics and City Science*, *45*(5), 817–820. DOI: 10.1177/2399808318796416

Batty, M. (2019). Urban analytics defined. *Environment and Planning. B, Urban Analytics and City Science*, *46*(3), 403–405. DOI: 10.1177/2399808319839494

Batty, M., Axhausen, K. W., Giannotti, F., Pozdnoukhov, A., Bazzani, A., Wachowicz, M., Ouzounis, G., & Portugali, Y. (2012). Smart cities of the future. *The European Physical Journal. Special Topics, 214*(1), 481–518. DOI: 10.1140/epjst/e2012-01703-3

Beil, C., & Kolbe, T. H. (2017). CityGML and the streets of New York: A proposal for detailed street space modelling. In *Proceedings of the 12th International 3D GeoInfo Conference 2017*, 9–16. DOI: 10.5194/isprs-annals-IV-4-W5-9-2017

Berlato, M., Binni, L., Durmus, D., Gatto, C., Giusti, L., Massari, A., Toldo, B. M., Cascone, S., & Mirarchi, C. (2025). Digital platforms for the built environment: A systematic review across sectors and scales. *Buildings (Basel, Switzerland), 15*(14), 2432. DOI: 10.3390/buildings15142432

Bhatt, S., Simre, S. S., Patadiya, H. H., Yadav, D., Deepti, B., Sathvi, S. S., & Manamasa, Y. (2025). 3D Printing in Dentistry: A Review. *Journal of Advanced Medical and Dental Sciences Research, 13*(1), 11–15.

Bianchini, S., Damioli, G., & Ghisetti, C. (2023). The environmental effects of the "twin" green and digital transition in European regions. *Environmental and Resource Economics, 84*(4), 877–918. Advance online publication. DOI: 10.1007/s10640-022-00741-7 PMID: 36532702

Biljecki, F., Stoter, J., Ledoux, H., Zlatanova, S., & Çöltekin, A. (2015). Applications of 3D City Models: State of the Art Review. *ISPRS International Journal of Geo-Information, 4*(4), 2842–2889. DOI: 10.3390/ijgi4042842

Boccardo, P., Riccia, L., & Yadav, Y. (2024). Urban echoes: Exploring the dynamic realities of cities through digital twins. *Land (Basel), 13*(5), 635. DOI: 10.3390/land13050635

Borkowski, A. S., & Olszewska, P. (2025). BIM model of district heating networks in design and investment management processes: A case study. *Sustainability (Basel), 17*(9), 4102. DOI: 10.3390/su17094102

Borkowski, A., Kochański, Ł., & Wyszomirski, M. (2022). A case study on building information (bim) and land information (lim) models including geospatial data. *Geomatics and Environmental Engineering, 17*(1), 19–34. DOI: 10.7494/geom.2023.17.1.19

Breunig, M., Bradley, P. E., Jahn, M., Kuper, P., Mazroob, N., Rösch, N., Al-Doori, M., Stefanakis, E., & Jadidi, M. (2020). Geospatial data management research: Progress and future directions. *ISPRS International Journal of Geo-Information, 9*(2), 95. Advance online publication. DOI: 10.3390/ijgi9020095

Bruno, N., Rechichi, F., Achille, C., Zerbi, A., Roncella, R., & Fassi, F. (2020). INTEGRATION OF HISTORICAL GIS DATA IN A HBIM SYSTEM. *The International Archives of the Photogrammetry, Remote Sensing and Spatial Information Sciences, XLIII-B4-2020*, 427–434. DOI: 10.5194/isprs-archives-XLIII-B4-2020-427-2020

Bryde, D., Broquetas, M., & Volm, J. M. (2013). The project benefits of Building Information Modelling (BIM). *International Journal of Project Management, 31*(7), 971–980. DOI: 10.1016/j.ijproman.2012.12.001

Bueno, R. E., Tetu Moysés, S., Reis Bueno, P. A., Jorge Moysés, S., De Carvalho, M. L., & Sottile França, B. H. (2013). Sustainable development and child health in the Curitiba metropolitan mesoregion, State of Paraná, Brazil. *Health & Place, 19*, 167–173. DOI: 10.1016/j.healthplace.2012.10.007 PMID: 23262454

Buuveibaatar, M., Lee, K., & Lee, W. (2022). Development of a Conceptual Data Model for 3D Geospatial Road Management Based on LandInfra Standard: A Case Study of Korea. *ISPRS International Journal of Geo-Information, 11*(5), 316. Advance online publication. DOI: 10.3390/ijgi11050316

Buyukdemircioglu, M., & Kocaman, S. (2020). Reconstruction and Efficient Visualization of Heterogeneous 3D City Models. *Remote Sensing (Basel), 12*(13), 2128. DOI: 10.3390/rs12132128

Cantarero-García, G., Mollá, M., & González-Lezcano, R. A. (2023). Smart citizen in architecture and landscape. Method design based on spatial intelligence and universal accessibility learning by students with intellectual disabilities. *Frontiers in Built Environment, 9*, 1094760. DOI: 10.3389/fbuil.2023.1094760

Cao, Y., Xu, C., Aziz, N., & Kamaruzzaman, S. (2023). Bim–gis integrated utilization in urban disaster management: The contributions, challenges, and future directions. *Remote Sensing (Basel), 15*(5), 1331. DOI: 10.3390/rs15051331

Cardoso, M. A., Silva, M. S., Coelho, S. T., Almeida, M. C., & Covas, D. I. C. (2012). Urban water infrastructure asset management – a structured approach in four water utilities. *Water Science and Technology, 66*(12), 2702–2711. DOI: 10.2166/wst.2012.509 PMID: 23109589

Carrasco, C. A., Lombillo, I., Sánchez-Espeso, J. M., & Balbás, F. J. (2024). A new approach to 3D facilities management in buildings using GIS and BIM integration: A case study application. *Applied Sciences (Basel, Switzerland), 14*(23), 11097. DOI: 10.3390/app142311097

Carrasco, C. A., Lombillo, I., Sánchez-Espeso, J. M., Blanco, H., & Boffill, Y. (2024). Methodology for 3D management of university faculties using integrated GIS and BIM models: A case study. *Buildings (Basel, Switzerland), 14*(11), 3547. DOI: 10.3390/buildings14113547

Castañeda, K., Sánchez, O., Herrera, R. F., Gómez-Cabrera, A., & Mejía, G. (2024). Building Information Modeling uses and complementary technologies in road projects: A systematic review. *Buildings (Basel, Switzerland), 14*(3), 563. DOI: 10.3390/buildings14030563

Castellanos-Ardila, J. P., Gallina, B., & Governatori, G. (2021). Compliance-aware engineering process plans: The case of space software engineering processes. *Artificial Intelligence and Law, 29*(4), 587–627. DOI: 10.1007/s10506-021-09285-5

Cepa, J. J., Alberti, M. G., Pavón, R. M., & Calvo, J. A. (2024). Integrating BIM and GIS for an existing infrastructure. *Applied Sciences (Basel, Switzerland), 14*(23), 10962. DOI: 10.3390/app142310962

Chen, T., & Li, Y. (2024). Building urban resilience: Lessons from the COVID-19 pandemic for future-proofing city infrastructure. Journal of Urban Management. https://doi.org/DOI: 10.1016/j.jum.2024.11.016

Chen, Z., Chen, L., Zhou, X., Huang, L., Sandanayake, M., & Yap, P. S. (2024). Recent technological advancements in BIM and LCA integration for sustainable construction: A review. *Sustainability (Basel), 16*(3), 1340. DOI: 10.3390/su16031340

Chib, S., Mehta, A. K., Selvakumar, P., Saravanan, A., Mishra, B. R., & Manjunath, T. C. (2025). Mental health challenges and solutions in high-pressure work environments. In Harriott, S. (Ed.), *Innovative approaches to managing conflict and change in diverse work environments* (pp. 247–272). IGI Global Scientific Publishing. DOI: 10.4018/979-8-3693-9556-1.ch011

Chiodelli, F., & Morpurgo, D. (2022). Placing the law: The socio-spatial impact of legal norms beyond mere compliance. *Transactions of the Institute of British Geographers, 47*(3), 725–740. DOI: 10.1111/tran.12530

Chong, H. Y., Ma, Q., Lai, J., & Liao, X. (2025). Achieving Sustainable Construction Safety Management: The Shift from Compliance to Intelligence via BIM–AI Convergence. *Sustainability (Basel), 17*(10), 4454. DOI: 10.3390/su17104454

Coetzee, S., Ivánová, I., Mitasova, H., & Brovelli, M. (2020). Open Geospatial Software and Data: A Review of the Current State and A Perspective into the Future. *ISPRS International Journal of Geo-Information, 9*(2), 90. DOI: 10.3390/ijgi9020090

Colucci, E., Iacono, E., Matrone, F., & Ventura, G. (2023). The development of a 2d/3d bim-gis web platform for planned maintenance of built and cultural heritage: the main10ance project. The International Archives of the Photogrammetry Remote Sensing and Spatial Information Sciences, XLVIII-M-2-2023, 433-439. DOI: 10.5194/isprs-archives-XLVIII-M-2-2023-433-2023

Dahiya, R., & Selvakumar, P. Vidyadhara, & Manjunath, T. C. (2025). Design and aerodynamics: Principles and advances. In M. Mellal (Ed.), Innovations and developments in unmanned aerial vehicles (pp. 145-170). IGI Global Scientific Publishing.

Dahiya, R., & Selvakumar, P. Vidyadhara, & Manjunath, T. C. (2025). Design and aerodynamics: Principles and advances. In M. Mellal (Ed.), *Innovations and Developments in Unmanned Aerial Vehicles* (pp. 145-170). IGI Global Scientific Publishing.

Dall'O', G., Zichi, A., & Torri, M. (2020). Green BIM and CIM: Sustainable Planning Using Building Information Modelling. In G. Dall'O' (Ed.), *Green Planning for Cities and Communities* (pp. 383–409). Springer International Publishing. DOI: 10.1007/978-3-030-41072-8_17

Dantas, H., Sousa, J., & Melo, C. (2019). The importance of city information modeling (CIM) for cities' sustainability. [IOP Publishing.]. *IOP Conference Series. Earth and Environmental Science*, *225*, 012074. DOI: 10.1088/1755-1315/225/1/012074

Darsin, M., Pamungkas, W. J., Syuhri, S. N. H., Wibowo, R. K. K., Basuki, H. A., Djumharianto, D., Yudistiro, D., Choiron, M. A., Wardana, M. K. A., & Thongchai, S. (2025). Advanced optimization of drone frame design through the application of generative design techniques and 3D printing technology. *Journal of Mechanical Science and Technology*, *39*(1), 1–10. DOI: 10.1007/s12206-024-1210-x

Das, D. K. (2024). Exploring the symbiotic relationship between digital transformation, infrastructure, service delivery, and governance for smart sustainable cities. *Smart Cities*, *7*(2), 806–835. DOI: 10.3390/smartcities7020034

de Vries, W. T. (2023). Ethical and Legal Issues Affecting Geospatial Sciences and Advances in Smart Land Management. In *Geospatial Science for Smart Land Management* (pp. 421–430). CRC Press. DOI: 10.1201/9781003349518-31

del Cerro Velázquez, F., & Morales Méndez, G. (2021). Systematic review of the development of spatial intelligence through augmented reality in stem knowledge areas. *Mathematics*, *9*(23), 3067. DOI: 10.3390/math9233067

Delbosc, A., & Currie, G. (2011). Using Lorenz curves to assess public transport equity. *Journal of Transport Geography*, *19*(6), 1252–1259. DOI: 10.1016/j.jtrangeo.2011.02.008

Demir, S., & Yomralioglu, T. (2024). Bridging geo-data and natural gas pipeline design standards: A systematic review of BIM-GIS integration for natural gas pipeline asset management. *Energies*, *17*(10), 2306. DOI: 10.3390/en17102306

Deng, Y., Cheng, J. C. P., & Anumba, C. (2016). A framework for 3D traffic noise mapping using data from BIM and GIS integration. *Structure and Infrastructure Engineering*, *12*(10), 1267–1280. DOI: 10.1080/15732479.2015.1110603

Deprêtre, A., Mielniczek, A., & Jacquinod, F. (2023). Le City Information Modelling (CIM) au service d'un projet urbain: Retour d'expérience sur la première phase de mise en œuvre du CIM d'un quartier. *Flux*, *3*(133), 57–75. DOI: 10.3917/flux1.133.0057

Di Giuda, G. M., Meschini, S., Gasbarri, P., Accardo, D., Tagliabue, L. C., & Cacciaguerra, E. (2024). BIM, GIS and BI tools for a university asset management system supporting space management, occupancy evaluation and optimization strategies. [ITcon]. *Information Technology and Control*, *29*, 1128–1155.

Ding, X., Yang, J., Liu, L., Huang, W., & Wu, P. (2020). Integrating ifc and citygml model at schema level by using linguistic and text mining techniques. *IEEE Access : Practical Innovations, Open Solutions*, *8*, 56429–56440. DOI: 10.1109/ACCESS.2020.2982044

Du, H. (2024). Urban Landscape Accessibility Evaluation Model based on GIS and spatial analysis. *Results in Engineering*, *103723*. Advance online publication. DOI: 10.1016/j.rineng.2024.103723

Du, J., Zou, Z., Shi, Y., & Zhao, D. (2018). Zero latency: Real-time synchronization of BIM data in virtual reality for collaborative decision-making. *Automation in Construction*, *85*, 51–64. DOI: 10.1016/j.autcon.2017.10.009

Durdyev, S., Ashour, M., Connelly, S., & Mahdiyar, A. (2022). Barriers to the implementation of Building Information Modelling (BIM) for facility management. *Journal of Building Engineering*, *46*, 103736. Advance online publication. DOI: 10.1016/j.jobe.2021.103736

DURSUN. İ., Varlık, A., & Ayyıldız, E. (2022). Integration and web-based presentation of 3d city models and bim data: the case of the köyceğiz campus. The International Archives of the Photogrammetry Remote Sensing and Spatial Information Sciences, XLIII-B4-2022, 513-519. https://doi.org/DOI: 10.5194/isprs-archives-xliii-b4-2022-513-2022

El Ammari, K., & Hammad, A. (2019). Remote interactive collaboration in facilities management using BIM-based mixed reality. *Automation in Construction*, *107*, 102940. DOI: 10.1016/j.autcon.2019.102940

ElBeheiry, D. M. (2025). An Integrated BIM and GIS Framework for Sustainable Urban Planning. *JES. Journal of Engineering Sciences*, *0*(0), 270–287. DOI: 10.21608/jesaun.2025.358173.1422

ElBeheiry, D. M. (2025). An integrated BIM and GIS framework for sustainable urban planning. *JES. Journal of Engineering Sciences*, *53*(3), 270–287.

Elgayar, A., & Jrade, A. (2022). Integrating 3D modeling and GIS-Based land use and Multimodal Transport Accessibility Evaluation model to design and retrofit sustainable bridges. In Lecture notes in civil engineering (pp. 83–93). https://doi.org/DOI: 10.1007/978-981-19-0968-9_7

Elshaboury, N., AlMetwaly, W. M., Hesham, A., & Abbas, A. (2025). Integrated BIM-GIS framework for holistic building stock assessment using 5D geo-modeling and digital twin concepts. *Journal of Building Engineering*, *111*, 113391. DOI: 10.1016/j.jobe.2025.113391

Elsheikh, A., Alzamili, H. H., Al-Zayadi, S. K., & Alboo-Hassan, A. S. (2021). Integration of GIS and BIM in Urban Planning -A Review. *IOP Conference Series. Materials Science and Engineering*, *1090*(1), 012128. DOI: 10.1088/1757-899X/1090/1/012128

Fan, C., Hou, J., Li, X., Song, G., Yang, Y., Liang, X., Zhou, Q., Imran, M., Chen, G., Wang, Z., & Lu, P. (2025). Efficient urban flood control and drainage management framework based on digital twin technology and optimization scheduling algorithm. *Water Research*, *282*, 123711. DOI: 10.1016/j.watres.2025.123711 PMID: 40319783

Fang, J., Xie, X., & Xu, J. (2024). Development and key technologies for architectural design and technical solutions of cim platform: A review. *IOP Conference Series. Earth and Environmental Science*, *1337*(1), 012036. DOI: 10.1088/1755-1315/1337/1/012036

Fang, X., Li, J., & Ma, Q. (2023). Integrating green infrastructure, ecosystem services and nature-based solutions for urban sustainability: A comprehensive literature review. *Sustainable Cities and Society, 98*, 104843. DOI: 10.1016/j.scs.2023.104843

Fernández-Rodríguez, S., Cortés-Pérez, J. P., Muriel, P. P., Tormo-Molina, R., & Maya-Manzano, J. M. (2018). Environmental impact assessment of Pinaceae airborne pollen and green infrastructure using BIM. *Automation in Construction, 96*, 494–507. DOI: 10.1016/j.autcon.2018.10.011

Furlong, C., Brotchie, R., Considine, R., Finlayson, G., & Guthrie, L. (2017). Key concepts for Integrated Urban Water Management infrastructure planning: Lessons from Melbourne. *Utilities Policy, 45*, 84–96. DOI: 10.1016/j.jup.2017.02.004

Gao, Y., Xiong, G., Hu, Z., Chai, C., & Li, H. (2024). Bridge digital twin for practical bridge operation and maintenance by integrating GIS and BIM. *Buildings (Basel, Switzerland), 14*(12), 3731. DOI: 10.3390/buildings14123731

García-Granero, Á., Jerí-McFarlane, S., Torres-Marí, N., Brogi, L., Ferrà-Canet, M., Navarro Zoroa, M. Á., Gamundí-Cuesta, M., & González-Argenté, F. X. (2025). 3D-reconstruction printed models and virtual reality improve teaching in oncological colorectal surgery. *Techniques in Coloproctology, 29*(1), 24. DOI: 10.1007/s10151-024-03074-3 PMID: 39699719

Gargiulo, I., Garcia, X., Benages-Albert, M., Martinez, J., Pfeffer, K., & Vall-Casas, P. (2020). Women's safety perception assessment in an urban stream corridor: Developing a safety map based on qualitative GIS. *Landscape and Urban Planning, 198*, 103779. Advance online publication. DOI: 10.1016/j.landurbplan.2020.103779

Geurs, K. T., Boon, W., & Van Wee, B. (2008). Social Impacts of Transport: Literature review and the state of the practice of transport appraisal in the Netherlands and the United Kingdom. *Transport Reviews, 29*(1), 69–90. DOI: 10.1080/01441640802130490

Gholami, M., Torreggiani, D., Tassinari, P., & Barbaresi, A. (2022). Developing a 3D City Digital Twin: Enhancing Walkability through a Green Pedestrian Network (GPN) in the City of Imola, Italy. *Land (Basel), 11*(11), 1917. Advance online publication. DOI: 10.3390/land11111917

Gijre, V., & Gupta, S. (2020). Urban transport governance practice and challenges in an Emerging Economy – Case study of India. *Transportation Research Procedia, 48*, 2435–2445. DOI: 10.1016/j.trpro.2020.08.293

Gikas, V. (2012). Three-dimensional laser scanning for geometry documentation and construction management of highway tunnels during excavation. *Sensors (Basel)*, *12*(8), 11249–11270. DOI: 10.3390/s120811249 PMID: 23112655

Gil, J., Almeida, J., & Duarte, J. P. (2011). The backbone of a city information model (CIM). In *Respecting Fragile Places: Education in Computer Aided Architectural Design in Europe*, 143–151.

Giraldo, J. P., Cho, S. S., Eghrari, N. B., Dholaria, N., Farber, S. H., Ehredt, R. B., ... & Uribe, J. S. (2025). Advances in neurosurgical education: Literature review of mixed-reality simulation models and novel mixed-reality spine prototype. Journal of Neurosurgery: Spine, 1(aop), 1-14.

Glander, T., & Döllner, J. (2009). Abstract representations for interactive visualization of virtual 3D city models. *Computers, Environment and Urban Systems*, *33*(5), 375–387. DOI: 10.1016/j.compenvurbsys.2009.07.003

Gong, Z., Chen, J., Jiao, X., Gong, H., Pan, D., Liu, L., Zhang, Y., & Tan, T. (2024). Genome-scale metabolic network models for industrial microorganisms metabolic engineering: Current advances and future prospects. *Biotechnology Advances*, *72*, 108319. DOI: 10.1016/j.biotechadv.2024.108319 PMID: 38280495

Gonthier, C. (2022). Cross-cultural differences in visuo-spatial processing and the culture-fairness of visuo-spatial intelligence tests: An integrative review and a model for matrices tasks. *Cognitive Research: Principles and Implications*, *7*(1), 11. DOI: 10.1186/s41235-021-00350-w PMID: 35119577

Goodchild, M. F., & Glennon, J. A. (2010). Crowdsourcing geographic information for disaster response: A research frontier. *International Journal of Digital Earth*, *3*(3), 231–241. DOI: 10.1080/17538941003759255

Graziano, A. D., Ragusa, E., Trifilò, D., Triaca, L. M., Trombetti, M., & Arcidiacono, C. (2023). Interaction Between Airport Facilities And The Surrounding Area Within The GIS And BIM Interoperability. *Transportation Research Procedia*, *69*, 273–280. DOI: 10.1016/j.trpro.2023.02.172

Green, D., O'Donnell, E., Johnson, M., Slater, L., Thorne, C., Zheng, S., Stirling, R., Chan, F. K. S., Li, L., & Boothroyd, R. J. (2021). Green infrastructure: The future of urban flood risk management? *WIREs. Water*, *8*(6), e1560. DOI: 10.1002/wat2.1560

Haidar, C. H., Maqsoom, A., Choudhary, S., Shahzad, M., & Khalid, M. H. (2024, December). Rainwater harvesting for urban flood mitigation: A BIM-GIS integrated approach. In *2024 International Conference on Frontiers of Information Technology (FIT)* (pp. 1-6). IEEE. DOI: 10.1109/FIT63703.2024.10838457

Hammad, A. W., Akbarnezhad, A., Haddad, A., & Vazquez, E. G. (2019). Sustainable Zoning, Land-Use Allocation and Facility Location Optimisation in Smart Cities. *Energies*, *12*(7), 1318. DOI: 10.3390/en12071318

Hanna, E., & Comín, F. A. (2021). Urban Green Infrastructure and Sustainable Development: A Review. *Sustainability (Basel)*, *13*(20), 11498. DOI: 10.3390/su132011498

Haveman, S., Peng, X., Zhang, R., & Rahola, T. (2024). Towards social digital twins – an integrated sociotechnical approach for the urban energy transition., 479-496. DOI: 10.18690/um.fov.4.2024.28

He, S., Zhu, D., Chen, Y., Liu, X., Chen, Y., & Wang, X. (2020). Application and problems of emergy evaluation: A systemic review based on bibliometric and content analysis methods. *Ecological Indicators*, *114*, 106304. DOI: 10.1016/j.ecolind.2020.106304

Hevko, I. V., Potapchuk, O. I., Lutsyk, I. B., Yavorska, V. V., Hiltay, L. S., & Stoliar, O. B. (2025). A novel pedagogical approach to equipping prospective IT professionals with skills in 3D modelling and reconstruction of architectural heritage. In *CEUR Workshop Proceedings* (pp. 85-101).

Hijazi, I., Ehlers, M., Zlatanova, S., Becker, T., & Berlo, L. (2010). Initial investigations for modeling interior utilities within 3d geo context: transforming ifc-interior utility to citygml/utilitynetworkade., 95-113. DOI: 10.1007/978-3-642-12670-3_6

Hijazi, I., Krauth, T., Donaubauer, A., & Kolbe, T. (2020). 3dcitydb4bim: a system architecture for linking bim server and 3d citydb for bim-gis-integration. Isprs Annals of the Photogrammetry Remote Sensing and Spatial Information Sciences, V-4-2020, 195-202. DOI: 10.5194/isprs-annals-V-4-2020-195-2020

Hogan, M., Ojo, A., Harney, O., Ruijer, E., Meijer, A., Andriessen, J., Pardijs, M., Boscolo, P., Palmisano, E., Satta, M., Groff, J., Baker, M., Détienne, F., Porwol, L., Scarano, V., & Malandrino, D. (2017). Governance, Transparency and the Collaborative Design of Open Data Collaboration Platforms: Understanding Barriers, Options, and Needs. In Ojo, A., & Millard, J. (Eds.), *Government 3.0 – Next Generation Government Technology Infrastructure and Services* (Vol. 32, pp. 299–332). Springer International Publishing., DOI: 10.1007/978-3-319-63743-3_12

Hor, A., Jadidi, A., & Sohn, G. (2016). Bim-gis integrated geospatial information model using semantic web and rdf graphs. *ISPRS Annals of the Photogrammetry, Remote Sensing and Spatial Information Sciences*, *III-4*, 73–79. DOI: 10.5194/isprs-annals-III-4-73-2016

Hou, Y., Heo, J., & Chang, S. (2023). Information visualization platform for underground pipeline maintenance integrating BIM and GIS data. In *Computing in Civil Engineering 2023* (pp. 781-789).

Huang, S., Tang, L., Hupy, J. P., Wang, Y., & Shao, G. (2021). A commentary review on the use of normalized difference vegetation index (NDVI) in the era of popular remote sensing. In *Journal of Forestry Research* (Vol. 32, Issue 1). DOI: 10.1007/s11676-020-01155-1

Huang, X., Wang, S., Lu, T., Liu, Y., & Serrano-Estrada, L. (2024). Crowdsourced geospatial data is reshaping urban sciences. *International Journal of Applied Earth Observation and Geoinformation*, *127*, 103687. DOI: 10.1016/j.jag.2024.103687

Hu, J., Lv, Z., Yuan, D., He, B., Chen, W., Ye, X., Li, D., & Yang, G. (2023). A spatiotemporal intelligent framework and experimental platform for urban digital twins. *Virtual Reality & Intelligent Hardware*, *5*(3), 213–231. DOI: 10.1016/j.vrih.2022.08.018

Hung, V. V., & Phuong, N. C. (2025). Integration of BIM and GIS on web viewer environment for Tan Cang station asset management. *Transportation Research Procedia*, *85*, 281–288. DOI: 10.1016/j.trpro.2025.03.163

Ilesanmi, A. O. (2010). Urban sustainability in the context of Lagos mega-city. *Journal of Geography and Regional Planning*, *3*(10), 240–252.

Irschick, D. J., Christiansen, F., Hammerschlag, N., Martin, J., Madsen, P. T., Wyneken, J., Brooks, A., Gleiss, A., Fossette, S., Siler, C., Gamble, T., Fish, F., Siebert, U., Patel, J., Xu, Z., Kalogerakis, E., Medina, J., Mukherji, A., Mandica, M., & Lauder, G. (2022). 3D visualization processes for recreating and studying organismal form. *iScience*, *25*(9), 104867. DOI: 10.1016/j.isci.2022.104867 PMID: 36060053

Isikay, I., Cekic, E., Baylarov, B., Tunc, O., & Hanalioglu, S. (2024). Narrative review of patient-specific 3D visualization and reality technologies in skull base neurosurgery: Enhancements in surgical training, planning, and navigation. *Frontiers in Surgery*, *11*, 1427844. DOI: 10.3389/fsurg.2024.1427844 PMID: 39081485

Islam, M. S., Yeasmin, T., Karmaker, S., Hossain, M. S., & Shi, L. (2023). Vegetation Cover Change Analysis during 1989-2020 of Coastal Barguna District, Bangladesh Using Remote Sensing and GIS Technology. *International Review for Spatial Planning and Sustainable Development*, *11*(2), 259–277. Advance online publication. DOI: 10.14246/irspsd.11.2_259

Ismaeil, E. M. (2024). Asset information model management-based GIS/BIM integration in facility management contract. *Sustainability (Basel), 16*(6), 2495. DOI: 10.3390/su16062495

Jain, M., Khan, S. A., Sharma, K., Jadhao, P. R., Pant, K. K., Ziora, Z. M., & Blaskovich, M. A. T. (2022). Current perspective of innovative strategies for bioremediation of organic pollutants from wastewater. *Bioresource Technology, 344*, 126305. DOI: 10.1016/j.biortech.2021.126305 PMID: 34752892

Jalaei, F. (2015). *Integrate Building Information Modeling (BIM) and Sustainable Design at the Conceptual Stage of Building Projects*. DOI: 10.20381/RUOR-4260

Jillella, S., Matan, A., & Newman, P. (2015). Participatory Sustainability Approach to Value Capture-Based Urban Rail Financing in India through Deliberated Stakeholder Engagement. *Sustainability (Basel), 7*(7), 8091–8115. DOI: 10.3390/su7078091

Jo, H., Lee, J. K., Lee, Y. C., & Choo, S. (2024). Generative artificial intelligence and building design: Early photorealistic render visualization of façades using local identity-trained models. *Journal of Computational Design and Engineering, 11*(2), 85–105. DOI: 10.1093/jcde/qwae017

Johansson, T., Segerstedt, E., Olofsson, T., & Jakobsson, M. (2016). Revealing Social Values by 3D City Visualization in City Transformations. *Sustainability (Basel), 8*(2), 195. DOI: 10.3390/su8020195

Jovanović, D., Milovanov, S., Ruskovski, I., Govedarica, M., Sladić, D., Radulović, A., & Pajić, V. (2020). Building virtual 3D city model for smart cities applications: A case study on campus area of the University of Novi Sad. *ISPRS International Journal of Geo-Information, 9*(8), 476. DOI: 10.3390/ijgi9080476

Kaewpoonsuk, P., & Subsomboo, K. (2024). Urban information modeling for smart cities: Advancing sustainable infrastructure and environmental management in digital twin applications. *Pakistan Journal of Life and Social Sciences, 22*(2). Advance online publication. DOI: 10.57239/PJLSS-2024-22.2.001545

Kampczyk, A., & Dybeł, K. (2021). The fundamental approach of the digital twin application in railway turnouts with innovative monitoring of weather conditions. *Sensors (Basel), 21*(17), 5757. Advance online publication. DOI: 10.3390/s21175757 PMID: 34502648

Kang, T. (2023). Scan to bim mapping process description for building representation in 3d gis. *Applied Sciences (Basel, Switzerland), 13*(17), 9986. DOI: 10.3390/app13179986

Kang, T. W., & Hong, C. H. (2015). A study on software architecture for effective BIM/GIS-based facility management data integration. *Automation in Construction*, *54*, 25–38. DOI: 10.1016/j.autcon.2015.03.019

Karner, A., London, J., Rowangould, D., & Manaugh, K. (2020). From transportation equity to transportation justice: Within, through, and beyond the state. *Journal of Planning Literature*, *35*(4), 440–459. DOI: 10.1177/0885412220927691

Kashyap, B., & Kumar, R. (2021). Sensing Methodologies in Agriculture for Soil Moisture and Nutrient Monitoring. *IEEE Access : Practical Innovations, Open Solutions*, *9*, 14095–14121. Advance online publication. DOI: 10.1109/ACCESS.2021.3052478

Kaur, M., Selvakumar, P., & Dinesh, N. Mohit, Chandel, P. S., & Manjunath, T. C. (2025). Reconstruction and preservation strategies for cultural heritage. In S. Oflazoğlu & M. Dora (Eds.), Preserving cultural heritage in post-disaster urban renewal (pp. 385-412). IGI Global Scientific Publishing.

Kaur, M., Selvakumar, P., & Dinesh, N. Mohit, Chandel, P. S., & Manjunath, T. C. (2025). Reconstruction and preservation strategies for cultural heritage. In *S. Oflazoğlu & M. Dora (Eds.), Preserving Cultural Heritage in Post-Disaster Urban Renewal* (pp. 385-412). IGI Global Scientific Publishing.

Keivani, R. (2009). A review of the main challenges to urban sustainability. *International Journal of Urban Sustainable Development*, *1*(1–2), 5–16. DOI: 10.1080/19463131003704213

Khoshamadi, N., Banihashemi, S., Poshdar, M., Abbasianjahromi, H., Tabadkani, A., & Hajirasouli, A. (2023). Parametric and generative mechanisms for infrastructure projects. *Automation in Construction*, *154*, 104968. DOI: 10.1016/j.autcon.2023.104968

Kiavarz, H., Jadidi, M., Rajabifard, A., & Sohn, G. (2018). Bim-gis oriented intelligent knowledge discovery. *The International Archives of the Photogrammetry, Remote Sensing and Spatial Information Sciences*, *XLII-4*(W10), 79–82. DOI: 10.5194/isprs-archives-XLII-4-W10-79-2018

Kommineni, M., Chundru, S., Maroju, P. K., & Selvakumar, P. (2025). Ethical implications of AI in sustainable development pedagogy. In Chinedu, C., Mafarja, N., Hussein, M., & Saleem, A. (Eds.), *Rethinking the pedagogy of sustainable development in the AI era* (pp. 17–36). IGI Global Scientific Publishing. DOI: 10.4018/979-8-3693-9062-7.ch002

Kujala, J., Heikkinen, A., Jokinen, A., & Tapaninaho, R. (2022). Stakeholder Engagement in the Generation of Urban Ecosystem Services: The Case of the Vuores Stormwater System. *South Asian Journal of Business and Management Cases*, *11*(1), 49–63. DOI: 10.1177/22779779221079484

Kumar, S. (2024b, December 18). Singapore Smart Traffic System is Redefining Urban Mobility. IntelliStride. https://www.intellistride.com/blog/how-singapores-smart-traffic-system-is-redefining-urban-mobility/

Kumar, K., Labetski, A., Ohori, K. A., Ledoux, H., & Stoter, J. (2019a). Harmonising the OGC standards for the built environment: A CityGML extension for Landinfra. *ISPRS International Journal of Geo-Information*, *8*(6), 246. Advance online publication. DOI: 10.3390/ijgi8060246

Kumar, K., Labetski, A., Ohori, K. A., Ledoux, H., & Stoter, J. (2019b). The LandInfra standard and its role in solving the BIM-GIS quagmire. *Open Geospatial Data. Software and Standards*, *4*(1), 5. Advance online publication. DOI: 10.1186/s40965-019-0065-z

Kutzner, T., Chaturvedi, K., & Kolbe, T. H. (2020). CityGML 3.0: New Functions Open Up New Applications. *PFG -. Journal of Photogrammetry, Remote Sensing and Geoinformation Science*, *88*(1), 43–61. Advance online publication. DOI: 10.1007/s41064-020-00095-z

Lai, J., Wan, R., Chong, H. Y., & Liao, X. (2025). Digital Intelligence in Building Lifecycle Management: A Mixed-Methods Approach. *Sustainability (Basel)*, *17*(11), 5121. DOI: 10.3390/su17115121

Lasorella, M., & Cantatore, E. (2025). CityGML-based model for the recovery and management of built cultural heritage: A systematic review towards digitalized processes. City. *Territory and Architecture*, *12*(1), 10. DOI: 10.1186/s40410-025-00259-7

Lawal, O. O., & Nawari, N. O. (2024). Blockchain-enabled city information modeling framework for urban asset management. *Journal of Architectural Engineering*, *30*(3), 04024020. DOI: 10.1061/JAEIED.AEENG-1644

Lee, C., & Yoon, Y. (2022). A novel bus arrival time prediction method based on Spatio-Temporal flow centrality analysis and deep learning. *Electronics (Basel)*, *11*(12), 1875. DOI: 10.3390/electronics11121875

Lee, D., & Lee, S. (2021). Digital twin for supply chain coordination in modular construction. *Applied Sciences (Basel, Switzerland)*, *11*(13), 5909. Advance online publication. DOI: 10.3390/app11135909

Lee, J., Lee, J. H., Im, J., & Jang, J. (2024). Enhancing urban safety: Optimal patrol route strategies for volunteer security squads based on integrated BIM-GIS data. *Buildings (Basel, Switzerland), 14*(10), 3074. DOI: 10.3390/buildings14103074

Lefranc, G., Cabrera, M. P., Comparan, R. O., & López-Juárez, I. (2025). Advanced decision-making strategies and technologies for manufacturing: Case studies, and future research directions. *International Journal of Computers, Communications & Control, 20*(1). Advance online publication. DOI: 10.15837/ijccc.2025.1.6906

Leng, S. and Zhen-Zhong, H. (2020). A lightweight bim-gis integration method for rural building design and construction., 25-30. DOI: 10.3311/CCC2020-071

Lewin, S. S. (2012). URBAN SUSTAINABILITY AND URBAN FORM METRICS. *Journal of Green Building, 7*(2), 44–63. DOI: 10.3992/jgb.7.2.44

Leyzerova, A., Sharovarova, E., & Alekhin, V. (2016). Sustainable Strategies of Urban Planning. *Procedia Engineering, 150*, 2055–2061. DOI: 10.1016/j.proeng.2016.07.299

Li, S., Suh, W., & Kim, J. I. (2024). An integrated urban road rehabilitation planning and decision support based on BIM and GIS. In *Computing in Civil Engineering 2023* (pp. 112-120). DOI: 10.1061/9780784485231.014

Liang, Y., Li, J., & Liu, H. (2021). GIS-based analysis of urban traffic model: Case study in Beijing. E3S Web of Conferences, 268, 01056. https://doi.org/DOI: 10.1051/e3sconf/202126801056

Li, D., Shao, Z., & Zhang, R. (2020). Advances of geo-spatial intelligence at LIESMARS. *Geo-Spatial Information Science, 23*(1), 40–51. DOI: 10.1080/10095020.2020.1718001

Limbore, N. V., Anute, N., Ahluwalia, G. K., Chordiya, P. B., Kalshetti, P., & Selvakumar, P. (2025). Ethical leadership and corporate responsibility. In Harriott, S. (Ed.), *Innovative Approaches to Managing Conflict and Change in Diverse Work Environments* (pp. 193–218). IGI Global Scientific Publishing. DOI: 10.4018/979-8-3693-9556-1.ch009

Li, N., Chan, D., Mao, Q., Hsu, K., & Fu, Z. (2018). Urban sustainability education: Challenges and pedagogical experiments. *Habitat International, 71*, 70–80. DOI: 10.1016/j.habitatint.2017.11.012

Lindgren, D. (2020). Framework for assessing compliance. In *An Assessment Framework for Compliance with International Space Law and Norms: Promoting Equitable Access and Use of Space for Emerging Actors* (pp. 49–84). Springer International Publishing. DOI: 10.1007/978-3-030-15762-3_5

Liu, J.-M., Setiazi, H., & So, P.-Y. (2023). Fisheries hydroacoustic assessment: A bibliometric analysis and direction for future research towards a blue economy. *Regional Studies in Marine Science, 60*, 102838. DOI: 10.1016/j.rsma.2023.102838

Liu, L., Li, B., Zlatanova, S., & van Oosterom, P. (2021). Indoor navigation supported by the Industry Foundation Classes (IFC): A survey. *Automation in Construction, 121*, 103436. DOI: 10.1016/j.autcon.2020.103436

Liu, R., Chen, Y., Wu, J., Xu, T., Gao, L., & Zhao, X. (2018). Mapping spatial accessibility of public transportation network in an urban area – A case study of Shanghai Hongqiao Transportation Hub. *Transportation Research Part D, Transport and Environment, 59*, 478–495. DOI: 10.1016/j.trd.2018.01.003

Liu, X., Payakkamas, P., Dijk, M., & De Kraker, J. (2023). GIS Models for Sustainable Urban Mobility Planning: Current use, future needs and Potentials. *Future Transportation, 3*(1), 384–402. DOI: 10.3390/futuretransp3010023

Liu, X., Wang, X., Wright, G., Cheng, J., Li, X., & Liu, R. (2017). A state-of-the-art review on the integration of building information modeling (bim) and geographic information system (gis). *ISPRS International Journal of Geo-Information, 6*(2), 53. DOI: 10.3390/ijgi6020053

Liu, Y., Liu, Y., Chen, X., Lin, C., & Zhou, Z. (2025). Balancing spatial logic and governance objectives for street vending: An optimizing model for dynamic formalization zones. *Cities (London, England), 160*, 105819. DOI: 10.1016/j.cities.2025.105819

Liu, Z., Xia, F., Guo, B., Leng, C., Zhang, E., Xu, L., Chen, X., & Zhu, P. (2025). Comprehensive evaluation of the ramification patterns of hepatic vascular anatomy based on three-dimensional visualization technology. *Updates in Surgery, 77*(8), 1–13. DOI: 10.1007/s13304-025-02064-w PMID: 39853656

Li, X., Wang, J., & Zhang, L. (2025). Gamifying data visualization in smart cities: Fostering citizen engagement in urban monitoring. *Advances in Engineering Technology Research, 13*(1), 151–151. DOI: 10.56028/aetr.13.1.151.2025

Lopes, H. S., Vidal, D. G., Cherif, N., Silva, L., & Remoaldo, P. C. (2025). Green infrastructure and its influence on urban heat island, heat risk, and air pollution: A case study of Porto (Portugal). *Journal of Environmental Management, 376*, 124446. DOI: 10.1016/j.jenvman.2025.124446 PMID: 39921953

Lorenzo-Eiroa, P. (2022). Space-environment commons: From big data survey to AI, to a post-capitalist blockchain zoning platform. In *Resilient Communities and the Peccioli Charter: Towards the Possibility of an Italian Charter for Resilient Communities* (pp. 161–172). Springer International Publishing. DOI: 10.1007/978-3-030-85847-6_17

Lubida, A., Veysipanah, M., Pilesjo, P., & Mansourian, A. (2019). LAND-USE PLANNING FOR SUSTAINABLE URBAN DEVELOPMENT IN AFRICA: A SPATIAL AND MULTI-OBJECTIVE OPTIMIZATION APPROACH. *Geodesy and Cartography (Vilnius)*, *45*(5), 1–15. DOI: 10.3846/gac.2019.6691

Luederitz, C., Lang, D. J., & Von Wehrden, H. (2013). A systematic review of guiding principles for sustainable urban neighborhood development. *Landscape and Urban Planning*, *118*, 40–52. DOI: 10.1016/j.landurbplan.2013.06.002

Madeira, R. N., Santos, P. A., Java, O., Priebe, T., Graça, E., Sarközi, E., Asprion, B., & Gomez, R. P. B. (2022). Towards Digital Twins for Multi-Sensor Land and Plant Monitoring. *Procedia Computer Science*, *210*(C), 45–52. Advance online publication. DOI: 10.1016/j.procs.2022.10.118

Maharjan, S., Ma, C., Singh, B., Kang, H., Orive, G., Yao, J., & Zhang, Y. S. (2024). Advanced 3D imaging and organoid bioprinting for biomedical research and therapeutic applications. *Advanced Drug Delivery Reviews*, *208*, 115237. DOI: 10.1016/j.addr.2024.115237 PMID: 38447931

Maltseva, I., Meng, T., Xiao, S., Cao, Y., & Zeng, L. (2024). An integrated review and analysis of urban building seismic disaster management based on bim-gis. E3s Web of Conferences, 474, 02013. DOI: 10.1051/e3sconf/202447402013

Marrero, M., Wojtasiewicz, M., Martínez-Rocamora, A., Solís-Guzmán, J., & Alba-Rodríguez, M. D. (2020). BIM-LCA Integration for the Environmental Impact Assessment of the Urbanization Process. *Sustainability (Basel)*, *12*(10), 4196. DOI: 10.3390/su12104196

Martin, A. S., & Freeland, S. (2021). The advent of artificial intelligence in space activities: New legal challenges. *Space Policy*, *55*, 101408. DOI: 10.1016/j.spacepol.2020.101408

Martino, N., Sanders, C. B., & Dej, E. (2024). The 'regulatory grey zone': Bylaw enforcement's governing of homelessness and space. *Policing and Society*, *34*(4), 305–320. DOI: 10.1080/10439463.2023.2263617

Marzouk, M., & Othman, A. (2020). Planning utility infrastructure requirements for smart cities using the integration between BIM and GIS. *Sustainable Cities and Society*, *57*, 102120. DOI: 10.1016/j.scs.2020.102120

Maskeliūnas, R., Maqsood, S., Vaškevičius, M., & Gelšvartas, J. (2025). Fusing LiDAR and photogrammetry for accurate 3D data: A hybrid approach. *Remote Sensing (Basel)*, *17*(3), 443. DOI: 10.3390/rs17030443

Matniyazov, Z., Adilov, Z., Khotamov, A., Elmurodov, S., Rasul-Zade, L., & Abdikhalilov, F. (2025). Integration of BIM and GIS technologies in modern urban planning: Challenges and prospects. *American Journal of Education and Learning*, *3*(3), 972–977.

Mekawy, M., Issa, M., & Perjons, E. (2021). Bim-gis integration for traffic safety in cities. Isprs Annals of the Photogrammetry Remote Sensing and Spatial Information Sciences, VIII-4/W2-2021, 161-166. DOI: 10.5194/isprs-annals-VIII-4-W2-2021-161-2021

Meschini, S., Accardo, D., Avena, M., Seghezzi, E., Tagliabue, L., & Giuda, G. (2022). Data integration through a bim-gis web platform for the management of diffused university assets.. DOI: 10.35490/EC3.2022.217

Meschini, S., Pellegrini, L., Locatelli, M., Accardo, D., Tagliabue, L. C., Di Giuda, G. M., & Avena, M. (2022). Toward cognitive digital twins using a BIM-GIS asset management system for a diffused university. *Frontiers in Built Environment*, *8*, 959475. Advance online publication. DOI: 10.3389/fbuil.2022.959475

Milojević, B., & Kuvač, I. (2023). Recognizing Principles of Integrated Urban Planning in Historical Development of the City: A Case Study of Banja Luka. *Journal of Urban History*, *49*(2), 290–308. DOI: 10.1177/00961442211030624

Misilmani, A. H., El-Bastawissi, I. Y., Ayad, H., & El Sayary, S. (2024). BIM-GIS integration: An innovative tool to enhance urban heritage management in the digital era. [APJ]. *Architecture and Planning Journal*, *30*(1), 3. DOI: 10.54729/2789-8547.1245

Monteiro, R., Ferreira, J., & Antunes, P. (2020). Green Infrastructure Planning Principles: An Integrated Literature Review. *Land (Basel)*, *9*(12), 525. DOI: 10.3390/land9120525

Mourato, S., Fernandez, P., Pereira, L. G., & Moreira, M. (2023). Assessing Vulnerability in Flood Prone Areas Using Analytic Hierarchy Process—Group Decision Making and Geographic Information System: A Case Study in Portugal. *Applied Sciences (Basel, Switzerland)*, *13*(8), 4915. Advance online publication. DOI: 10.3390/app13084915

Mujib, M. A., Apriyanto, B., Kurnianto, F. A., Ikhsan, F. A., Nurdin, E. A., Pangastuti, E. I., & Astutik, S. (2021). Assessment of Flood Hazard Mapping Based on Analytical Hierarchy Process (AHP) and GIS: Application in Kencong District, Jember Regency, Indonesia. *Geosfera Indonesia*, *6*(3), 353. Advance online publication. DOI: 10.19184/geosi.v6i3.21668

Mushtaha, A. W., Alaloul, W. S., Musarat, M. A., Baarimah, A. O., Rabah, F. K., & Alawag, A. M. (2024). BIM-GIS Integration for Infrastructure Management in Post-Disaster Stage. In *2024 ASU International Conference in Emerging Technologies for Sustainability and Intelligent Systems (ICETSIS)* (pp. 856-861). IEEE. DOI: 10.1109/ICETSIS61505.2024.10459527

Mutuku, B., Boerboom, L., & Madureira, A. M. (2019). The role of Planning Support Systems in national policy transfer and policy translation in secondary cities. *International Planning Studies*, *24*(3–4), 293–307. DOI: 10.1080/13563475.2019.1657809

Nair, S. M., Andrushia, D., Carmichael, M. J., Ebenezer, J. S., & Varghese, A. (2025). GIS and BIM integration for assessing and maintaining existing buildings: a review. *Proceedings of the Institution of Civil Engineers-Forensic Engineering*.

Najafi, P., Mohammadi, M., van Wesemael, P., & Le Blanc, P. M. (2023). A user-centred virtual city information model for inclusive community design: State of art. *Cities (London, England)*, *134*, 104203. DOI: 10.1016/j.cities.2023.104203

Narindri, B., Nugroho, A., & Aminullah, A. (2022). Developing building management system framework using web-based-gis and bim integration. *Civil Engineering Dimension*, *24*(2), 71–84. DOI: 10.9744/ced.24.2.71-84

Nasr-Azadani, E., Wardrop, D., & Brooks, R. (2022). Is the rapid development of visualization techniques enhancing the quality of public participation in natural resource policy and management? A systematic review. *Landscape and Urban Planning*, *228*, 104586. DOI: 10.1016/j.landurbplan.2022.104586

National Geospatial Policy. 2022. (2022). https://www.pib.gov.in/PressReleasePage.aspx?PRID=2106569

Nchang, L. N., Fombe, L. F., & Wanie, C. M. (2024). Population vulnerability to the state of public transport in Bamenda city, Cameroon. *ISAR Journal of Arts. Humanities and Social Sciences*, *2*(10), 85–98.

Newell, R., McCarthy, N., Picketts, I., Davis, F., Hovem, G., & Navarrete, S. (2021). Communicating complexity: Interactive model explorers and immersive visualizations as tools for local planning and community engagement. *Facets*, *6*, 287–316. DOI: 10.1139/facets-2020-0045

Nica, E., Popescu, G. H., Poliak, M., Kliestik, T., & Sabie, O. M. (2023). Digital twin simulation tools, spatial cognition algorithms, and multi-sensor fusion technology in sustainable urban governance networks. *Mathematics*, *11*(9), 1981. DOI: 10.3390/math11091981

Nijkamp, P., Kourtit, K., Krugman, P., & Moreno, C. (2024). Old wisdom and the new economic geography: Managing uncertainty in 21st century regional and urban development. *Regional Science Policy & Practice*, *16*(10), 100124. DOI: 10.1016/j.rspp.2024.100124

Norrman, K. E. (2023). World population growth: A once and future global concern. *WORLD (Oakland, Calif.)*, *4*(4), 684–697. DOI: 10.3390/world4040043

Ohori, K., Diakité, A., Krijnen, T., Ledoux, H., & Stoter, J. (2018). Processing bim and gis models in practice: Experiences and recommendations from a geobim project in the netherlands. *ISPRS International Journal of Geo-Information*, *7*(8), 311. DOI: 10.3390/ijgi7080311

Okonta, E. D., Vukovic, V., & Hayat, E. (2024). Prospective Directions in the Computer Systems Industry Foundation Classes (IFC) for Shaping Data Exchange in the Sustainability and Resilience of Cities. *Electronics 2024, Vol. 13, Page 2297*, *13*(12), 2297. DOI: 10.3390/electronics13122297

Olanrewaju, O. I., Kineber, A. F., Chileshe, N., & Edwards, D. J. (2022). Modelling the relationship between Building Information Modelling (BIM) implementation barriers, usage and awareness on building project lifecycle. *Building and Environment*, *207*, 108556. Advance online publication. DOI: 10.1016/j.buildenv.2021.108556

Önden, İ., Pamucar, D., Deveci, M., As, Y., Birol, B., & Şüheda Yıldız, F. (2024). Prioritization of transfer centers using GIS and fuzzy Dombi Bonferroni weighted Assessment (DOBAS) model. *Expert Systems with Applications*, *238*, 121827. Advance online publication. DOI: 10.1016/j.eswa.2023.121827

Onungwa, I., Olugu-Uduma, N., & Shelden, D. (2021). Cloud bim technology as a means of collaboration and project integration in smart cities. *SAGE Open*, *11*(3), 21582440211033250. Advance online publication. DOI: 10.1177/21582440211033250

Open Geospatial Consortium. (2017). *InfraGML.* LandInfra Core - Encoding Standard.

Ouhajjou, N., Loibl, W., Fenz, S., & Tjoa, A. M. (2017). Stakeholder-oriented energy planning support in cities. *Sustainable Cities and Society*, *28*, 482–492. DOI: 10.1016/j.scs.2016.08.006

Ozel, B., & Petrovic, M. (2024). Green Urban Scenarios: A Framework for Digital Twin Representation and Simulation for Urban Forests and Their Impact Analysis. *Arboriculture & Urban Forestry*, *50*(1), jauf.2023.029. Advance online publication. DOI: 10.48044/jauf.2023.029

Panahi Rizi, M. H., & Hosseini Seno, S. A. (2022). A systematic review of technologies and solutions to improve security and privacy protection of citizens in the smart city. *Internet of Things : Engineering Cyber Physical Human Systems*, *20*, 100584. DOI: 10.1016/j.iot.2022.100584

Pangbourne, K., Mladenović, M. N., Stead, D., & Milakis, D. (2020). Questioning mobility as a service: Unanticipated implications for society and governance. *Transportation Research Part A, Policy and Practice*, *131*, 35–49. DOI: 10.1016/j.tra.2019.09.033

Papa, E., & Bertolini, L. (2015). Accessibility and transit-oriented development in European metropolitan areas. *Journal of Transport Geography*, *47*, 70–83. DOI: 10.1016/j.jtrangeo.2015.07.003

Paul, S. (2018). Geospatial world. *Urban Planning and GIS*. https://geospatialworld.net/blogs/gis-in-urban-planning/

Pavelka, K., Matoušková, E., & Pavelka, K.Jr. (2019). THE CONTRIBUTION OF GEOMATIC TECHNOLOGIES TO BIM. *The International Archives of the Photogrammetry, Remote Sensing and Spatial Information Sciences*, *XLII-5*(W3), 85–89. DOI: 10.5194/isprs-archives-XLII-5-W3-85-2019

Pelden, S., Banihashemi, S., Mohandes, S. R., Arashpour, M., & Kalantari, M. (2025). Enhancing infrastructure planning and design through BIM-GIS integration. *Structure and Infrastructure Engineering*, •••, 1–20. DOI: 10.1080/15732479.2025.2461629

Petojević, Z., Parezanović, A., Nadaždi, A., Isailović, D., & Višnjevac, N. (2025). Integration of BIM and GIS to reduce construction and demolition waste. In *Advances in Construction and Demolition Waste Recycling* (pp. 99–118). Woodhead Publishing. DOI: 10.1016/B978-0-443-23962-5.00002-5

Piras, G., Agostinelli, S., & Muzi, F. (2024). Digital twin framework for built environment: A review of key enablers. *Energies, 17*(2), 436. DOI: 10.3390/en17020436

Piras, G., Muzi, F., & Zylka, C. (2024). Integration of BIM and GIS for the digitization of the built environment. *Applied Sciences (Basel, Switzerland), 14*(23), 11171. DOI: 10.3390/app142311171

Pojani, D., & Stead, D. (2016). The urban transport crisis in emerging economies: An introduction. *Urban Transport in the Developing World*, 1–10. https://doi.org/ DOI: 10.1007/978-3-319-43851-1_1

Popoola, O. A. M., Carruthers, D., Lad, C., Bright, V. B., Mead, M. I., Stettler, M. E. J., Saffell, J. R., & Jones, R. L. (2018). Use of networks of low cost air quality sensors to quantify air quality in urban settings. *Atmospheric Environment, 194*, 58–70. Advance online publication. DOI: 10.1016/j.atmosenv.2018.09.030

Popov, P., Mozaffari, M. H., RazaviAlavi, S. R., & Jalaei, F. (2025). GIS-based solutions for management of public building and infrastructure assets: A review of state of the art and research trend analysis. *Smart Construction, 2*(2), 1–2. DOI: 10.55092/sc20250009

Postert, P., Wolf, A. E. M., & Schiewe, J. (2022). Integrating Visualization and Interaction Tools for Enhancing Collaboration in Different Public Participation Settings. *ISPRS International Journal of Geo-Information, 11*(3), 156. DOI: 10.3390/ijgi11030156

Qiu, S., Xie, B., Liu, Q., & Heng, P. A. (2025, January). Advancing extended reality with 3D Gaussian splatting: Innovations and prospects. In *2025 IEEE International Conference on Artificial Intelligence and eXtended and Virtual Reality (AIxVR)* (pp. 203-208). IEEE. DOI: 10.1109/AIxVR63409.2025.00039

Qiu, Y., Zuo, S., Yu, Z., Zhan, Y., & Ren, Y. (2021). Discovering the effects of integrated green space air regulation on human health: A bibliometric and meta-analysis. *Ecological Indicators, 132*, 108292. DOI: 10.1016/j.ecolind.2021.108292

Qi, Y., Liu, C., Scaioni, M., Li, Y., Qiao, Y., Ma, X., Wu, H., Zhang, K., & Wang, D. (2024). Geometric information constraint 3D object detection from LiDAR point cloud for autonomous vehicles under adverse weather. *Transportation Research Part C, Emerging Technologies, 161*, 104555. DOI: 10.1016/j.trc.2024.104555

Rahman, M., & Szabó, G. (2022). Sustainable Urban Land-Use Optimization Using GIS-Based Multicriteria Decision-Making (GIS-MCDM) Approach. *ISPRS International Journal of Geo-Information, 11*(5), 313. DOI: 10.3390/ijgi11050313

Ran, J., & Nedovic-Budic, Z. (2016). Integrating spatial planning and flood risk management: A new conceptual framework for the spatially integrated policy infrastructure. *Computers, Environment and Urban Systems, 57*, 68–79. DOI: 10.1016/j.compenvurbsys.2016.01.008

Rasheed, M. W., Tang, J., Sarwar, A., Shah, S., Saddique, N., Khan, M. U., Imran Khan, M., Nawaz, S., Shamshiri, R. R., Aziz, M., & Sultan, M. (2022). Soil Moisture Measuring Techniques and Factors Affecting the Moisture Dynamics: A Comprehensive Review. In *Sustainability (Switzerland)* (Vol. 14, Issue 18). DOI: 10.3390/su141811538

Razinkina, I., Bulatenko, M., Chernov, S., & Prasolov, V. (2022). Ethical and legal balance of modern economic intelligence. *The Journal of World Intellectual Property, 25*(2), 335–346. DOI: 10.1111/jwip.12225

Rezvani, S. M. H. S., Falcão, M. J., Komljenovic, D., & de Almeida, N. M. (2023). A systematic literature review on urban resilience enabled with asset and disaster risk management approaches and GIS-based decision support tools. *Applied Sciences (Basel, Switzerland), 13*(4), 2223. DOI: 10.3390/app13042223

Rodrigues, F., Teixeira, J., Matos, R., & Rodrigues, H. (2019). Development of a Web Application for Historical Building Management through BIM Technology. *Advances in Civil Engineering, 2019*(1), 9872736. DOI: 10.1155/2019/9872736

Rodríguez-Espinosa, V. M., Aguilera-Benavente, F., & Gómez-Delgado, M. (2020). Green infrastructure design using GIS and spatial analysis: A proposal for the Henares Corridor (Madrid-Guadalajara, Spain). *Landscape Research, 45*(1), 26–43. Advance online publication. DOI: 10.1080/01426397.2019.1569221

Rotilio, M., Marchionni, C., Massari, A., Di Giovanni, G., & Simeone, D. (2024, June). Efficiency management of built environment: The digitalization of energy and service grids. In *2024 IEEE International Workshop on Metrology for Living Environment (MetroLivEnv)* (pp. 355-360). IEEE. DOI: 10.1109/MetroLivEnv60384.2024.10615404

Saccucci, M., & Pelliccio, A. (2018). Integrated BIM-GIS system for the enhancement of urban heritage. *2018 Metrology for Archaeology and Cultural Heritage (MetroArchaeo)*, 222–226. DOI: 10.1109/MetroArchaeo43810.2018.13625

Sachdeva, R., Armstrong, A. K., Arnaout, R., Grosse-Wortmann, L., Han, B. K., Mertens, L., Moore, R. A., Olivieri, L. J., Parthiban, A., & Powell, A. J. (2024). Novel techniques in imaging congenital heart disease: JACC scientific statement. *Journal of the American College of Cardiology, 83*(1), 63–81. DOI: 10.1016/j.jacc.2023.10.025 PMID: 38171712

Saifi, S., & Anandakumar, R. M. (2025). Web-based visualization and rendering of aerial LiDAR point cloud for urban flood simulation. *International Journal of Disaster Resilience in the Built Environment, 16*(2), 260–274. DOI: 10.1108/IJDRBE-07-2023-0079

Saksena, H. T. (2024). Enhancing Indonesia's transportation infrastructure planning through BIM and digital twins. *FSTPT International Journal of Transportation Studies, 1*(2), 36–48.

Samal, A., & Misra, L. I. Mohit, Pougajendy, S., Chandel, P. S., Manjunath, T. C., & Selvakumar, P. (2025). Musealization and tourism practices. In S. Oflazoğlu & M. Dora (Eds.), Preserving cultural heritage in post-disaster urban renewal (pp. 115-142). IGI Global Scientific Publishing.

Samal, A., & Misra, L. I. Mohit, Pougajendy, S., Chandel, P. S., Manjunath, T. C., & Selvakumar, P. (2025). Musealization and tourism practices. In *S. Oflazoğlu & M. Dora (Eds.), Preserving Cultural Heritage in Post-Disaster Urban Renewal* (pp. 115-142). IGI Global Scientific Publishing.

Sani, M., Musliman, I., & Rahman, A. (2022). Ifc to citygml conversion algorithm based on geometry and semantic mapping. The International Archives of the Photogrammetry Remote Sensing and Spatial Information Sciences, XLVI-4/W3-2021, 287-293. DOI: 10.5194/isprs-archives-XLVI-4-W3-2021-287-2022

Sarigul, F. H., & Gunaydin, H. M. (2025). *Integrated BIM, GIS and interoperable digital technologies in lifecycle management of building construction projects: systematic literature review*. Smart and Sustainable Built Environment.

Saunders, M. J., & Da Silva, A. N. R. (2009). Reducing Urban Transport Energy dependence: A new Urban Development Framework and GIS-Based tool. *International Journal of Sustainable Transportation, 3*(2), 71–87. DOI: 10.1080/15568310701648037

Schaller, J. (1992). GIS application in environmental planning and assessment. *Computers, Environment and Urban Systems, 16*(4), 337–353. DOI: 10.1016/0198-9715(92)90015-J

Schiller, P. L., & Kenworthy, J. R. (2017). An introduction to sustainable transportation. In Routledge eBooks. https://doi.org/DOI: 10.4324/9781315644486

Schindler, M., Dionisio, R., & Kingham, S. (2020). Challenges of Spatial Decision-Support Tools in Urban Planning: Lessons from New Zealand's Cities. *Journal of Urban Planning and Development, 146*(2), 04020012. DOI: 10.1061/(ASCE)UP.1943-5444.0000575

Seif, H. G., & Hu, X. (2016). Autonomous driving in the iCity—HD maps as a key challenge of the automotive industry. *Engineering (Beijing), 2*(2), 159–162. DOI: 10.1016/J.ENG.2016.02.010

Senhorelo, A. P., Sousa, E. F., de, , Santos, A. R., dos, , Ferrari, J. L., Peluzio, J. B. E., & Carvalho, R. (2024). Application of Path Analysis and Remote Sensing to Assess the Interrelationships between Meteorological Variables and Vegetation Indices in the State of Espírito Santo, Southeastern Brazil. *Diversity, 16*(2), 90. Advance online publication. DOI: 10.3390/d16020090

Shaikh, M., & Birajdar, F. (2024). Advancements in remote sensing and GIS for sustainable groundwater monitoring: Applications, challenges, and future directions. International Journal of Research in Engineering. *Science and Management, 7*(3), 16–24.

Shamaee, S. H., Yousefi, H., & Zahedi, R. (2024). Assessing urban development indicators for environmental sustainability. *Discover Sustainability, 5*(1), 341. DOI: 10.1007/s43621-024-00563-1

Sharma, P., Singh, R., & Srivastava, A. (2021). Analyzing the role of geospatial technology in smart city development. In The urban book series (pp. 1–20). https://doi.org/DOI: 10.1007/978-3-030-71945-6_1

Sharma, A., Lysenko, A., Jia, S., Boroevich, K. A., & Tsunoda, T. (2024). Advances in AI and machine learning for predictive medicine. *Journal of Human Genetics, 69*(10), 487–497. DOI: 10.1038/s10038-024-01231-y PMID: 38424184

Sharma, N., Kumar, R., Vihari, N. S., Arora, M., & Saini, J. R. (2025). Prioritizing smart city themes for multi-national enterprises and United Nations Sustainable Development Goals. *Sustainability (Basel), 17*(10), 4251. DOI: 10.3390/su17104251

Sheina, S., Chubarova, K., Dementeev, D., & Kalitkin, A. (2023). Integration of BIM and GIS Technologies for Sustainable Development of the Construction Industry. In Guda, A. (Ed.), *Networked Control Systems for Connected and Automated Vehicles* (Vol. 509, pp. 1303–1311). Springer International Publishing., DOI: 10.1007/978-3-031-11058-0_132

Sheth, A., & Kochut, K. J. (1998). Workflow Applications to Research Agenda: Scalable and Dynamic Work Coordination and Collaboration Systems. In Doğaç, A., Kalinichenko, L., Özsu, M. T., & Sheth, A. (Eds.), *Workflow Management Systems and Interoperability* (pp. 35–60). Springer Berlin Heidelberg., DOI: 10.1007/978-3-642-58908-9_3

Shi, W., Kong, Q., & Liu, Y. (2008). A GPS/GIS Integrated System for Urban Traffic Flow Analysis. In 2008 11th International IEEE Conference on Intelligent Transportation Systems (pp. 844–849). IEEE. https://doi.org/DOI: 10.1109/ITSC.2008.4732569

Shiode, N. (2000). 3D urban models: Recent developments in the digital modelling of urban environments in three-dimensions. *GeoJournal*, *52*(3), 263–269. DOI: 10.1023/A:1014276309416

Siddall, E., Grey, T., & Dyer, M. (2013). Indicators and stakeholder engagement: A Dublin case study. *Proceedings of the Institution of Civil Engineers. Engineering Sustainability*, *166*(2), 85–97. DOI: 10.1680/ensu.12.00004

Šimenić, D. (2021). Building Information Modeling (BIM) for road infrastructure: TEM requirements and recommendations. Genève (Suisse), 119 p.

Singh, S. K., & Khubani, P. (2018). Traffic analysis of grade intersections and measures of congestion mitigation at Indore. *Pyrex Journal of Engineering and Manufacturing Technology*, 3(2), 10–19. https://www.researchgate.net/publication/325381098

Singh, S. K., Alasali, B. E., & Belbase, Y. (2025). Urban Transportation planning. In *Advances in computational intelligence and robotics*. (pp. 469–504). *IGI Global Scientific Publsihing*. https://doi.org/DOI: 10.4018/979-8-3693-7250-0.ch017

Singh, J., Chaudhary, H. K., & Malik, A. (2022). A Case Study on Indore BRTS with Reference to Other Indian Cities. *International Journal for Research in Applied Science and Engineering Technology*, *10*(10), 196–203. DOI: 10.22214/ijraset.2022.46976

Singh, S. K., & Bansal, A. (2021). Critical analysis and feasibility of PRT for Gurgaon City. [IJSRSET]. *International Journal of Scientific Research in Science, Engineering and Technology*, *8*(1), 83–86. DOI: 10.32628/IJSRSET21812

Singh, S. K., Jain, R., Palaniappan, D., & Parmar, K.. (2025). *Spatiotemporal analysis and machine learning-based prediction of air quality in Indian urban*. Environmental Research and Technology. DOI: 10.35208/ert.1587308

Singh, S. K., & Saraswat, A. (2019). Traffic simulation & integration using SUMO Simulator. *14th International Conference on Recent Trends in Engineering, Applied Science & Management (ICRTESM-19)*, 327–332. ISBN: 978-93-87793-71-2

Sivakumar, A., Melina, J. V. T., & Suresh, E. S. M. (2025). BIM-GIS integration for efficient management of urban underground stormwater networks. *Proceedings of the Institution of Civil Engineers-Smart Infrastructure and Construction*, 1-14. DOI: 10.1680/jsmic.23.00092

Sladić, G., Milosavljević, B., Nikolic, S., Sladić, D., & Radulović, A. (2021). A Blockchain Solution for Securing Real Property Transactions: A Case Study for Serbia. *ISPRS International Journal of Geo-Information*, *10*(1), 35. DOI: 10.3390/ijgi10010035

Sohaib, M., Najeeb, A., Umair, M., Khan, M. A., Zubair, M. U., Jehan, Z., & Khattak, A. (2024). Improving urban road infrastructure analysis and design using an integrated BIM-GIS and traffic microsimulation framework. *Innovative Infrastructure Solutions*, *9*(7), 285. DOI: 10.1007/s41062-024-01609-z

Solé-Ribalta, A., Gómez, S., & Arenas, A. (2016). A model to identify urban traffic congestion hotspots in complex networks. *Royal Society Open Science*, *3*(10), 160098. DOI: 10.1098/rsos.160098 PMID: 27853535

Song, Y., Wang, X., Tan, Y., Wu, P., Sutrisna, M., Cheng, J., & Hampson, K. (2017). Trends and opportunities of bim-gis integration in the architecture, engineering and construction industry: A review from a spatio-temporal statistical perspective. *ISPRS International Journal of Geo-Information*, *6*(12), 397. DOI: 10.3390/ijgi6120397

Soundariya, A., Behera, N. K., Selvakumar, P., Chandel, P. S., Gupta, S., & Manjunath, T. C. (2025). AI and environmental ethics: The role of AI in addressing climate change ethical considerations in AI-driven environmental solutions. In Chakraborty, S. (Ed.), *Human Values, Ethics, and Dignity in the Age of Artificial Intelligence* (pp. 321–346). IGI Global Scientific Publishing. DOI: 10.4018/979-8-3693-9035-1.ch013

Souza, L., & Bueno, C. (2022). City information modelling as a support decision tool for planning and management of cities: A systematic literature review and bibliometric analysis. *Building and Environment*, *207*, 108403. DOI: 10.1016/j.buildenv.2021.108403

Sperling, J. B., & Ramaswami, A. (2018). Cities and "budget-based" management of the energy-water-climate nexus: Case studies in transportation policy, infrastructure systems, and urban utility risk management. *Environmental Progress & Sustainable Energy*, *37*(1), 91–107. DOI: 10.1002/ep.12765

Sreekanthaswamy, N., Anitha, S., Singh, A., Jayadeva, S. M., Gupta, S., Manjunath, T. C., & Selvakumar, P. (2025). Digital tools and methods. In Çela, E., Paliktzoglou, V., Sengar, S., Eappen, P., & Vajjhala, N. (Eds.), *Enhancing school counseling with technology and case studies* (pp. 25–48). IGI Global Scientific Publishing. DOI: 10.4018/979-8-3693-8392-6.ch002

Sudhira, H. S., & Ramachandra, T. V. (2009). A Spatial Planning Support System for Managing Bangalore's Urban Sprawl. In Geertman, S., & Stillwell, J. (Eds.), *Planning Support Systems Best Practice and New Methods* (Vol. 95, pp. 175–190). Springer Netherlands., DOI: 10.1007/978-1-4020-8952-7_9

Sullivan, R. (2022). The Geopolitics of Urban Digital Twins: Spatial Cognition and Predictive Modeling Algorithms, Simulation and Virtualization Technologies, and Blockchain-enabled Cyber-Physical Systems. *Geopolitics, History, and International Relations*, *14*(2), 25–39. DOI: 10.22381/GHIR14220222

Sun, Y., & Li, S. (2016). Real-time collaborative GIS: A technological review. *ISPRS Journal of Photogrammetry and Remote Sensing*, *115*, 143–152. DOI: 10.1016/j.isprsjprs.2015.09.011

Taheri, A., & Sobanjo, J. (2024). Civil Integrated Management (CIM) for advanced level applications to transportation infrastructure: A state-of-the-art review. *Infrastructures*, *9*(6), 90. DOI: 10.3390/infrastructures9060090

Takva, Ç., Top, S. M., Gökgöz, B. İ., Gebel, Ş., İlerisoy, Z. Y., İlcan, H., & Şahmaran, M. (2024). Applicability of 3D concrete printing technology in building construction with different architectural design decisions in housing. *Journal of Building Engineering*, *98*, 111257. DOI: 10.1016/j.jobe.2024.111257

Tamilarasi, C., Sowmiya, K., Vignesh, G. D., Jasmine, R. L., Nehru, P., & Selvakumar, P. (2025). AI in renewable energy education: Teaching the future of sustainable power systems. In Chinedu, C., Mafarja, N., Hussein, M., & Saleem, A. (Eds.), *Rethinking the Pedagogy of Sustainable Development in the AI Era* (pp. 133–152). IGI Global Scientific Publishing. DOI: 10.4018/979-8-3693-9062-7.ch007

Tan, T., Mills, G., Papadonikolaki, E., & Liu, Z. (2021). Combining multi-criteria decision making (MCDM) methods with building information modelling (BIM): A review. *Automation in Construction*, *121*, 103451. DOI: 10.1016/j.autcon.2020.103451

Tan, Y., Liang, Y., & Zhu, J. (2023). Citygml in the integration of bim and the gis: Challenges and opportunities. *Buildings (Basel, Switzerland)*, *13*(7), 1758. DOI: 10.3390/buildings13071758

Tao, F., Qi, Q., Wang, L., & Nee, A. (2019). Digital Twins and Cyber–Physical Systems toward Smart Manufacturing and Industry 4.0: Correlation and Comparison. *Engineering (Beijing)*, *5*(4), 653–661. DOI: 10.1016/j.eng.2019.01.014

Terzis, D. (2022). Monitoring innovation metrics in construction and civil engineering: Trends, drivers and laggards. *Developments in the Built Environment*, *9*, 100064. DOI: 10.1016/j.dibe.2021.100064

Thakur, J. K., Singh, S. K., & Ekanthalu, V. S. (2017). Integrating remote sensing, geographic information systems and global positioning system techniques with hydrological modeling. In *Applied Water Science* (Vol. 7, Issue 4). DOI: 10.1007/s13201-016-0384-5

Thakuriah, P., Tilahun, N., & Zellner, M. (2016). Seeing cities through big data. In Springer geography. https://doi.org/DOI: 10.1007/978-3-319-40902-3

Tobiáš, P. (2016). Bim, gis and semantic models of cultural heritage buildings. *Geoinformatics Fce Ctu*, *15*(2), 27–42. DOI: 10.14311/gi.15.2.3

Todeschi, V., Pappalardo, S. E., Zanetti, C., Peroni, F., & De Marchi, M. (2022). Climate Justice in the City: Mapping Heat-Related Risk for Climate Change Mitigation of the Urban and Peri-Urban Area of Padua (Italy). *ISPRS International Journal of Geo-Information*, *11*(9), 490. Advance online publication. DOI: 10.3390/ijgi11090490

Tripathi, I., Froese, T. M., & Mallory-Hill, S. (2023). Applicability of BIM-IoT-GIS integrated digital twins for post occupancy evaluations. *Frontiers in Built Environment*, *9*, 1103743. Advance online publication. DOI: 10.3389/fbuil.2023.1103743

Tsilimantou, E., Delegou, E. T., Nikitakos, I. A., Ioannidis, C., & Moropoulou, A. (2020). GIS and BIM as Integrated Digital Environments for Modeling and Monitoring of Historic Buildings. *Applied Sciences (Basel, Switzerland)*, *10*(3), 1078. DOI: 10.3390/app10031078

Tunio, I. A., Memon, S. A., Abbasi, H., Abro, A. A., Kumar, L., Tunio, M. A., Tunio, A. R., & Kumar, J. (2025). Spatiotemporal Assessment of Land Use Land Cover Change Detection and Normalized Difference Vegetation Index in Desert Area. *Rangeland Ecology and Management*, *101*, 126–139. DOI: 10.1016/j.rama.2025.03.008

Uddin, M. K. S. (2024). A review of utilizing natural language processing and AI for advanced data visualization in real-time analytics. *Global Mainstream Journal*, *1*(4), 10–62304. DOI: 10.62304/ijmisds.v1i04.185

Ülger, M., & Kocaman, H. (2023). A new modeling approach with integrated use of BIM and GIS in rural resettlement areas. Available at *SSRN* 5006381.

Ullah, M., Bibi, A., Wahab, A., Hamayun, S., Rehman, M. U., Khan, S. U., Awan, U. A., Riaz, N., Naeem, M., Saeed, S., & Hussain, T. (2024). Shaping the future of cardiovascular disease by 3D printing applications in stent technology and its clinical outcomes. *Current Problems in Cardiology*, *49*(1), 102039. DOI: 10.1016/j.cpcardiol.2023.102039 PMID: 37598773

Usmani, A., Jadidi, M., & Sohn, G. (2021). Towards the automatic ontology generation and alignment of bim and gis data formats. Isprs Annals of the Photogrammetry Remote Sensing and Spatial Information Sciences, VIII-4/W2-2021, 183-188. DOI: 10.5194/isprs-annals-VIII-4-W2-2021-183-2021

Vacca, G., Quaquero, E., Pili, D., & Brandolini, M. (2018). Integrating bim and gis data to support the management of large building stocks. *The International Archives of the Photogrammetry, Remote Sensing and Spatial Information Sciences*, *XLII-4*, 647–653. DOI: 10.5194/isprs-archives-XLII-4-647-2018

Vaidyanathan, B., Maya, H. A., Xavier, S., & Kappanayil, M. (2025). 3-Dimensional virtual reality-based visualization of fetal cardiac anatomy with the use of ultrasound datasets. *JACC. Case Reports*, *30*(8), 103211. DOI: 10.1016/j.jaccas.2024.103211 PMID: 40250906

Van Tan, N., Le Vin, T., Van Do, T., & Thao, N. D. (2025, May). Integration of advanced surveying technologies into BIM-GIS systems of port infrastructure. In *Proceedings of 7th International Conference on Civil Engineering and Architecture,* Volume 1*: Proceedings of ICCEA 2024* (Vol. 640, p. 178). Springer Nature.

Vergara-Perucich, J. F. (2025). AI-Driven Deconstruction of Urban Regulatory Frameworks: Unveiling Social Sustainability Gaps in Santiago's Communal Zoning. *Urban Science (Basel, Switzerland)*, *9*(6), 186. DOI: 10.3390/urbansci9060186

Verma, P., & Raghubanshi, A. S. (2018). Urban sustainability indicators: Challenges and opportunities. *Ecological Indicators*, *93*, 282–291. DOI: 10.1016/j.ecolind.2018.05.007

Vignali, V., Acerra, E. M., Lantieri, C., Di Vincenzo, F., Piacentini, G., & Pancaldi, S. (2021). Building information Modelling (BIM) application for an existing road infrastructure. *Automation in Construction*, *128*, 103752. Advance online publication. DOI: 10.1016/j.autcon.2021.103752

Wang, J., Mo, F., Qiao, S., Feng, H., & Lyu, Z. (2025). Spatial computing in digital twins. *Digital Twin*, 2508268.

Wang, B. (2021). The seductive smart city and the benevolent role of transparency. [s]. *Interaction Design and Architecture*, (48), 100–121. DOI: 10.55612/s-5002-048-005

Wang, H., Pan, Y., & Luo, X. (2019). Integration of BIM and GIS in sustainable built environment: A review and bibliometric analysis. *Automation in Construction*, *103*, 41–52. DOI: 10.1016/j.autcon.2019.03.005

Wang, K. (2021). A theoretical analysis method of spatial analytic geometry and mathematics under digital twins. *Advances in Civil Engineering*, *2021*(1), 8910274. DOI: 10.1155/2021/8910274

Wang, M., Deng, Y., Won, J., & Cheng, J. C. P. (2019). An integrated underground utility management and decision support based on BIM and GIS. *Automation in Construction*, *107*, 102931. DOI: 10.1016/j.autcon.2019.102931

Wang, P., Rui, X., Gu, J., Huang, K., Zhou, L., & Jiang, M. (2025). Fast parametric modeling of visualized simulation and design for tracked vehicle system. *Advances in Engineering Software*, *201*, 103852. DOI: 10.1016/j.advengsoft.2024.103852

Wang, X., & Xie, M. (2024). Integration of 3DGIS and BIM and its application in visual detection of concealed facilities. *Geo-Spatial Information Science*, *27*(1), 132–141. DOI: 10.1080/10095020.2022.2054732

Wang, X., Yang, J., Han, J., Wang, W., & Wang, F. Y. (2022). Metaverses and DeMetaverses: From digital twins in CPS to parallel intelligence in CPSS. *IEEE Intelligent Systems*, *37*(4), 97–102. DOI: 10.1109/MIS.2022.3196592

Wang, Y., Wang, Z., Zhao, D., Yu, X., Cheng, T., Bao, G., & Wang, Z. L. (2021). Flow and level sensing by waveform coupled liquid-solid contact-electrification. *Materials Today Physics*, *18*, 100372. Advance online publication. DOI: 10.1016/j.mtphys.2021.100372

Wang, Y., & Yang, Y. (2021). Analyzing the green innovation practices based on sustainability performance indicators: A Chinese manufacturing industry case. *Environmental Science and Pollution Research International*, *28*(1), 1181–1203. DOI: 10.1007/s11356-020-10531-7 PMID: 32833175

Wu, D., Zheng, P., Zhao, Q., Zhang, S., Qi, J., Hu, J., & Wang, L. (2026). Empowering natural human–robot collaboration through multimodal language models and spatial intelligence: Pathways and perspectives. *Robotics and Computer-integrated Manufacturing*, *97*, 103064. DOI: 10.1016/j.rcim.2025.103064

Wu, T., Yuan, Y. J., Zhang, L. X., Yang, J., Cao, Y. P., Yan, L. Q., & Gao, L. (2024). Recent advances in 3D Gaussian splatting. *Computational Visual Media*, *10*(4), 613–642. DOI: 10.1007/s41095-024-0436-y

Xhafa, S., & Kosovrasti, A. (2015). Geographic Information Systems (GIS) in urban planning. *European Journal of Interdisciplinary Studies*, *1*(1), 85. DOI: 10.26417/ejis.v1i1.p85-92

Xia, H., Liu, Z., Efremochkina, M., Liu, X., & Lin, C. (2022). Study on city digital twin technologies for sustainable smart city design: A review and bibliometric analysis of geographic information system and building information modeling integration. *Sustainable Cities and Society*, *84*, 104009. DOI: 10.1016/j.scs.2022.104009

Xie, Y., Liu, C., Chang, S., & Jiang, B. (2022). Urban Sustainability: Integrating Socioeconomic and Environmental Data for Multi-Objective Assessment. *Sustainability (Basel)*, *14*(15), 9142. DOI: 10.3390/su14159142

Xie, Y., Liu, J., Hu, N., Li, D., Song, Z., & Zhang, Z. (2025). Towards Ethical Spatial Decision-Making in Geospatial Artificial Intelligence. In *GeoAI and Human Geography: The Dawn of a New Spatial Intelligence Era* (pp. 313–325). Springer Nature Switzerland. DOI: 10.1007/978-3-031-87421-5_22

Xu, H., Omitaomu, F., Sabri, S., Zlatanova, S., Li, X., & Song, Y. (2024). Leveraging generative AI for urban digital twins: A scoping review on the autonomous generation of urban data, scenarios, designs, and 3D city models for smart city advancement. *Urban Informatics*, *3*(1), 29. DOI: 10.1007/s44212-024-00060-w

Xu, X., Ding, L., Luo, H., & Ma, L. (2014). From building information modeling to city information modeling. [ITcon]. *Journal of Information Technology in Construction*, *19*, 292–307.

Xu, Z., & Coors, V. (2012). Combining system dynamics model, GIS and 3D visualization in sustainability assessment of urban residential development. *Building and Environment*, *47*, 272–287. DOI: 10.1016/j.buildenv.2011.07.012

Xu, Z., Qi, M., Wu, Y., Hao, X., & Yang, Y. (2021). City information modeling: State of the art. *Applied Sciences (Basel, Switzerland)*, *11*(19), 9333. DOI: 10.3390/app11199333

Yamamura, S., Fan, L., & Suzuki, Y. (2017). Assessment of Urban Energy Performance through Integration of BIM and GIS for Smart City Planning. *Procedia Engineering*, *180*, 1462–1472. DOI: 10.1016/j.proeng.2017.04.309

Yan, S., & Tang, J. (2021). Optimization of green space planning to improve ecosystem services efficiency: The case of chongqing urban areas. *International Journal of Environmental Research and Public Health*, *18*(16), 8441. Advance online publication. DOI: 10.3390/ijerph18168441 PMID: 34444194

Yao, J., Murray, A. T., Wang, J., & Zhang, X. (2019). Evaluation and development of sustainable urban land use plans through spatial optimization. *Transactions in GIS*, *23*(4), 705–725. DOI: 10.1111/tgis.12531

Yau, S. S., Pandya, K., & Choudhary, S. (2024, July). Regulatory Compliance in Software Services Using Emerging Technologies. In *2024 IEEE International Conference on Software Services Engineering (SSE)* (pp. 36-42). IEEE. DOI: 10.1109/SSE62657.2024.00018

Yenisetty, P. T., & Bahadure, P. (2021). Assessing accessibility to ASFs from bus stops using distance measures: Case of two Indian cities. *Land Use Policy*, *108*, 105567. DOI: 10.1016/j.landusepol.2021.105567

Yiğit, A. Y., & Uysal, M. (2025). Virtual reality visualisation of automatic crack detection for bridge inspection from 3D digital twin generated by UAV photogrammetry. *Measurement*, *242*, 115931. DOI: 10.1016/j.measurement.2024.115931

Yigitcanlar, T., & Teriman, S. (2015). Rethinking sustainable urban development: Towards an integrated planning and development process. *International Journal of Environmental Science and Technology*, *12*(1), 341–352. DOI: 10.1007/s13762-013-0491-x

Yu, W., Zhou, X., Wang, D., & Dong, J. (2025). The development and construction of City Information Modeling (CIM): A survey from data perspective. *Applied Sciences (Basel, Switzerland)*, *15*(9), 4696. DOI: 10.3390/app15094696

Zaman, N., Ong, J., Waisberg, E., Masalkhi, M., Lee, A. G., Tavakkoli, A., & Zuckerbrod, S. (2024). Advanced visualization engineering for vision disorders: A clinically focused guide to current technology and future applications. *Annals of Biomedical Engineering*, *52*(2), 178–207. DOI: 10.1007/s10439-023-03379-8 PMID: 37861913

Zaroujtaghi, A., Mansourihanis, O., Tayarani, M., Mansouri, F., Hemmati, M., & Soltani, A. (2025). *A Systematic Review of GIS Evolution in Transportation Planning: Towards AI Integration*. MDPI. DOI: 10.3390/futuretransp5030097

Zartha Sossa, J. W., Gaviria Suárez, J. F., López Suárez, N. M., Rebolledo, J. L. S., Orozco Mendoza, G. L., & Vélez Suárez, V. (2022). Innovation Systems and Sustainability. Development of a Methodology on Innovation Systems for the Measurement of Sustainability Indicators in Regions Based on a Colombian Case Study. *Sustainability (Basel)*, *14*(23), 15955. DOI: 10.3390/su142315955

Zhang, H., & Srinivasan, R. (2020). A systematic review of air quality sensors, guidelines, and measurement studies for indoor air quality management. In *Sustainability (Switzerland)* (Vol. 12, Issue 21). DOI: 10.3390/su12219045

Zhang, L., Li, Y., Pan, Y., & Ding, L. (2024). Advanced informatic technologies for intelligent construction: A review. *Engineering Applications of Artificial Intelligence*, *137*, 109104. DOI: 10.1016/j.engappai.2024.109104

Zhao, T., Song, C., Yu, J., Xing, L., Xu, F., Li, W., & Wang, Z. (2025). Leveraging Immersive Digital Twins and AI-Driven Decision Support Systems for Sustainable Water Reserves Management: A Conceptual Framework. *Sustainability 2025, Vol. 17, Page 3754*, *17*(8), 3754. DOI: 10.3390/su17083754

Zhao, L., Liu, Z., & Mbachu, J. (2019). An integrated bim–gis method for planning of water distribution system. *ISPRS International Journal of Geo-Information*, *8*(8), 331. DOI: 10.3390/ijgi8080331

Zhao, L., Liu, Z., & Mbachu, J. (2019). Highway alignment optimization: An integrated bim and gis approach. *ISPRS International Journal of Geo-Information*, *8*(4), 172. DOI: 10.3390/ijgi8040172

Zheng, X., Zhang, S., Lin, W., Zhang, A., Mayol-Cuevas, W. W., & Shen, J. (2025). CULTURE3D: Cultural landmarks and terrain dataset for 3D applications. arXiv preprint arXiv:2501.06927.

Zhou, Y., & Liu, J. (2024). Advances in emerging digital technologies for energy efficiency and energy integration in smart cities. *Energy and Building*, *315*, 114289. DOI: 10.1016/j.enbuild.2024.114289

Zhu, J., Wright, G., Wang, J., & Wang, X. (2018). A critical review of the integration of geographic information system and building information modelling at the data level. *ISPRS International Journal of Geo-Information*, *7*(2), 66. DOI: 10.3390/ijgi7020066

Zhu, J., & Wu, P. (2021). Towards effective bim/gis data integration for smart city by integrating computer graphics technique. *Remote Sensing (Basel)*, *13*(10), 1889. DOI: 10.3390/rs13101889

Zhu, J., Wu, P., & Anumba, C. (2021). A semantics-based approach for simplifying ifc building models to facilitate the use of bim models in gis. *Remote Sensing (Basel)*, *13*(22), 4727. DOI: 10.3390/rs13224727

Zhu, J., Wu, P., Chen, M., Kim, M., Wang, X., & Fang, T. (2020). Automatically processing ifc clipping representation for bim and gis integration at the process level. *Applied Sciences (Basel, Switzerland)*, *10*(6), 2009. DOI: 10.3390/app10062009

About the Contributors

Munir Ahmad is a seasoned professional in the realm of Spatial Data Infrastructure (SDI), Geo-Information Productions, Information Systems, and Information Governance, boasting over 25 years of dedicated experience in the field. With a PhD in Computer Science, Dr. Ahmad's expertise spans Spatial Data Production, Management, Processing, Analysis, Visualization, and Quality Control. Throughout his career, Dr. Ahmad has been deeply involved in the development and deployment of SDI systems specially in the context of Pakistan, leveraging his proficiency in Spatial Database Design, Web, Mobile & Desktop GIS, and Geo Web Services Architecture. His contributions to Volunteered Geographic Information (VGI) and Open Source Geoportal & Metadata Portal have significantly enriched the geospatial community. As a trainer and researcher, Dr. Ahmad has authored over 50 publications, advancing the industry's knowledge base and fostering innovation in Geo-Tech, Data Governance, and Information Infrastructure, and Emerging Technologies. His commitment to Research and Development (R&D) is evident in his role as a dedicated educator and mentor in the field.

Jaishri Gothania is an academician, author, researcher, and data scientist with over 9 years of experience. She has published 12 research articles, presented 6 conference papers, and holds multiple patents. Her work bridges data science and innovation, contributing significantly to academia, research, and technological advancement.

S. Ida Evangeline, presently working as an Assistant Professor in the Department of Electrical and Electronics Engineering at the Government College of Engineering, Tirunelveli, Tamilnadu, India. She obtained her Bachelor's in Electrical and Electronics Engineering from Manonmaniam Sundaranar University in 2003, and Master's in Power Systems Engineering from Anna University, Chennai, Tamilnadu, India in 2005. She completed her PhD degree in 2023 at Anna University, Chennai, Tamilnadu, India. She has published six research papers in international journals. Her current research interests include optimal power flow, renewable energy, soft computing, machine learning, and deep learning.

Leevesh Kumar-Associate Professor Civil Engineering Dr. Leevesh Kumar is a distinguished academician and researcher with over 13 years of progressive experience in the field of civil and construction engineering across India, Ethiopia, the United Kingdom, and Bangladesh. Currently serving as an associate professor at the World University of Bangladesh and holding a concurrent part-time role with Liverpool John Moores University and the University of East London (UK), Dr. Kumar has built a diverse and global portfolio in higher education and construction technology research. He holds a Ph.D. in Civil Engineering from Career Point University (India), preceded by both M.Tech and B.Tech degrees in Civil Engineering from Bhagwant University. His educational foundation is complemented by a postgraduate diploma and several certifications from prestigious institutions like IIT Madras and NITTTR Chandigarh. Throughout his academic journey, Dr. Kumar has remained committed to advancing sustainable practices in the construction industry. Dr. Kumar's research focuses on sustainable construction materials, with specific interest in the mechanical and durability

P. Selvakumar is currently an Associate Professor in the Department of Science and Humanities at Nehru Institute of Technology, Coimbatore, Tamil Nadu, India. Prior to joining Nehru Institute of Technology, he held positions as an Assistant Professor in the Department of Science and Humanities various engineering colleges in Coimbatore, Tamil Nadu, India, spanning from 2011 to 2023. Dr.P. Selvakumar's academic and research interests lie primarily in the field of phytochemistry. He has made significant contributions to his field, with an impressive record that includes 30 journal publications, 12 patents, and 20 conference papers. Additionally, he has organized 5 seminars and conferences, further demonstrating his commitment to academic and scientific exchange. Throughout his career, Dr. P.Selvakumar has been recognized for his dedication and excellence in teaching, having been honored with several awards, including the Best Teacher Awards. He holds a prominent position in the academic community, contributing actively to research.

Ravikumar R. Natarajan is an experienced academician with over 12+ years of teaching and research expertise across institutions in India, the UAE, and the Maldives. He currently serves as an Assistant Professor at Marwadi University, Rajkot, specializing in Machine Learning, Recommender Systems, and AI-driven applications in healthcare and security. He holds a Ph.D. in Machine Learning and Recommender Systems from Amity University, Jaipur, a Master of Engineering in CSE from VMRF University, Salem, and a Bachelor of Engineering in CSE from

Anna University. He also has a MiniMBA from the Swiss eLearning Institute. Dr. Ravikumar has authored more than 8+ multiple SCIE/Scopus-indexed journal articles, 25+ IEEE/Springer conference papers, and co-authored a book on Artificial Intelligence. He has also published 9 patents and contributed 10+ book chapters in emerging domains such as Federated Learning, Blockchain, and Digital Twins for IGI Global, Wiley and Springer publisher. He actively mentors students through initiatives like the Research Activities Club (RAC) and Student Research Internship Training (SRIT), promoting innovation and research culture.

Jyoti Rani is an accomplished Assistant Professor at the School of Fashion Design and Technology, with over two years of experience in academia and research. Her work focuses on innovative and sustainable practices in fashion and textiles, making her a prominent figure in the field. Her research areas include vegan leather, natural dyeing, consumer behavior, artificial intelligence (AI) in fashion, and merchandising, reflecting her commitment to addressing contemporary challenges in the industry. Dr. Rani has published 28 research papers in reputed, peer-reviewed journals, all indexed in Scopus, demonstrating the high quality and impact of her work. Additionally, she has contributed 10 book chapters, showcasing her interdisciplinary expertise and ability to connect traditional practices with modern advancements. Her research often bridges the gap between sustainability and technology, particularly in areas like vegan leather and natural dyeing, which are critical to reducing the environmental footprint of the fashion industry.One of her standout achievements is securing 1 patent in fiber technology, highlighting her innovative approach to material science and its applications

Aarthi S is a dedicated Assistant Professor in Computer Science with over six years of teaching experience. She is currently pursuing her Ph.D. at Marwadi University, Gujarat, and holds an M.Phil. and MCA from Bharathiar University, Tamil Nadu. Throughout her academic career, Aarthi has specialized in programming languages such as C++, Java, and C, and contributed to research in recommender systems and AI-related data privacy. She has published journal articles, contributed to book chapters, and actively participated in technical workshops. Known for her student mentorship and engagement in academic events, Aarthi remains committed to fostering innovation and excellence in computer science education.

Sitesh Kumar Singh is an accomplished academic and Chartered Engineer with a Ph.D. in Civil Engineering and over nine years of diverse experience in education, research, consultancy, and educational administration. Currently serving as an Assistant Professor in the Department of Civil and Environmental Engineering at the National University of Science and Technology, Muscat, Sultanate of Oman, he specializes in Civil and Transportation Engineering, Project Management, and Data Analytics. Dr. Singh holds multiple advanced degrees including M.Tech, B.Tech, PGDIPR, PGDUPDL, and PGDEd. His career encompasses curriculum development, research supervision, quality assurance, and strategic academic planning. He has supervised over 40 postgraduate and 2 doctoral students, contributing to academic leadership through workshops, conferences, and editorial roles, and has published over 40 research articles, several patents, and authored books and book chapters.

Index

Symbols

B

C

D

G

I

L

M

P

www.ingramcontent.com/pod-product-compliance
Lightning Source LLC
LaVergne TN
LVHW081521300926
849161LV00006B/1257